Economics of the Family and Farming Systems in Sub-Saharan Africa

Economics of the Family and Farming Systems in Sub-Saharan Africa

Development Perspectives

Ram D. Singh

Foreword by Theodore W. Schultz

Westview Press
BOULDER & LONDON

Westview Special Studies in Social, Political, and Economic Development

This Westview softcover edition is printed on acid-free paper and bound in softcovers that carry the highest rating of the National Association of State Textbook Administrators, in consultation with the Association of American Publishers and the Book Manufacturers' Institute.

Published in 1988 in the United States of America by Westview Press, Inc., 5500 Central Avenue, Boulder, Colorado 80301

Library of Congress Cataloging-in-Publication Data
Singh, Ram D.
Economics of the family and farming systems in sub-Saharan Africa.
(Westview special studies in social, political, and economic development)
Includes index.
1. Family farms—Africa, Sub-Saharan. I. Title.
II. Series.
HD1476.A357S56 1988 338.1'0967 88-10729
ISBN 0-8133-7624-6

Printed and bound in the United States of America

The paper used in this publication meets the requirements of the American National Standard for Permanence of Paper for Printed Library Materials Z39.48-1984.

6 5 4 3 2 1

Dedicated to
the hardworking and intelligent farmers
of Burkina Faso

Contents

Tables

Foreword

It has long been my contention that all too little is known about families, households and farms throughout the poorest parts of Africa. Although much has been written and said about their plight, most of it has been superficial and largely wrong for reasons of not knowing how very poor African people manage to survive despite their adverse circumstances and meager options.

This book offers an important advance in basic knowledge of the behavior of these poor people. From it we learn how they manage, not by hit and miss, but to a fine degree their meager resources in production and in consumption. We also learn how families, households and farms are organized to cope with the existing bad conditions and the reasons why these people do what we observe them to do. This is what Professor Singh's research is all about.

The idea of poverty in high income countries is not at issue in this study. The reason for that is clear. The age-old concept of being poor is more comprehensive and useful in searching for dependable information than the now-politicized concept of poverty. Being poor is the lot of most people throughout the world. Africa has far more than its share of them. Among the poorest of the poor are the people in Burkina Faso, in western sub-Saharan Africa. At the time Professor and Mrs. Singh took up residence there to do this study, it was known as Upper Volta. The new name is less harsh; it has a fine ring.

Singh's sharp, inquiring mind is clearly evident. His apprenticeship could not have been better when it came to doing his pioneering work in Burkina Faso during 1979 through 1981. He benefitted from his earlier research on the economic lot of small farmers in India, which culminated in a decade of successful work as a member of the faculty of India's leading agricultural university. He also had a year with the Ford Foundation on rural issues in Chile. He has published landmark research on the lack of schooling and other forms of human capital in low income areas of Brazil. Had he concentrated all this time on farm families in Iowa, he would have been far less able to do what he succeeded in doing in Burkina Faso.

Not least is Professor Singh's knowledge of recent advances in the economics of the family, household economics and the economics of human capital. Readers of this book about Burkina Faso will benefit greatly from Singh's long and relevant experience in the field and from his professional competence. The problems he illuminates are of utmost importance to the people of the world; I hope this research will be widely read and understood.

Theodore W. Schultz

Preface

The book presents an integrated analysis of the dynamics of the economics of households and farm production in the setting of a polygynous family structure and traditional agriculture in sub-Saharan rural Africa. The section on the economics of the family provides indepth but at the same time simple explanations of the otherwise intricate decision making process of farm families, in particular the decisions about polygyny, the number of children, the allocation of time in home production and farm work, and the household's labor force participation in the external labor market. For the first time, estimates of the economic contributions of wives and children through home production and the dollar value of work at home have been made for an African LDC, indeed for any LDC. The section on farming systems presents a critical evaluation of the economics of the farm by providing an insight understanding of the existing farm production system, the pattern of resource allocation on the farm, yields under the current (traditional) crop patterns, the yield potential of some of the new production technologies, the labor productivity differentials and the effects of schooling on farm production and incomes.

Furthermore, this section highlights the constraints facing the farmers, and it also indicates some major problems confronted by the national and international institutions engaged in research and development in the region. Overall, an attempt has been made to synthesize the economics of the family and the farm, and to focus on some basic facts about poverty among the farm people of rural Africa. I hope this book adds some important dimensions to what Professor T.W. Schultz calls the "Economics of Being Poor," and provides a useful perspective on development.

I have dedicated this work to the farm people of Burkina Faso, once called the Upper Volta, in western sub-Saharan Africa. The work is an outcome of my association with a regional research and development project in Burkina Faso as an economist and team leader of the Farming Systems Research Project undertaken by Purdue University (1979-81) with the financial and logistic support of the U.S. Agency for International Development (USAID). I must confess that the more I learned about the farm people and the prevailing farming conditions, the more inadequate and at the same time curious I felt about the complexities and uniqueness of the African socio-economic systems. I always remained conscious of my limitations. However, this consciousness prompted me to involve myself more and more deeply to study the region's economy.

The results of findings reported in this volume are based upon (a) household survey data gathered from three sample regions of Burkina Faso and (b) personal notes and observations compiled during two years of my stay in the country. Information gathering in the study region involved a number of channels and processes: for example, interviewing sample farmers by trained field investigators conversant in the local conditions; field trials and demonstrations in farmers' fields, crop and soil surveys, crop cuttings for production estimates, labor time use studies; personal group discussions and visits to select numbers of farmers in the sample village by the author; interviewing school teachers and village chiefs; and data collection also from official records. There were approximately 25 to 30 field investigators whose survey work was supervised. The field staff lived in the villages during the entire period of investigation that covered one full agricultural year, in addition to the three months of reconnaissance survey during which the questionnaires were tested and sample selections made by the project team.

A large number of individuals and institutions were associated with the project--the source of data for this book, and I must acknowledge with gratitude their contributions.

The first and the foremost to thank are indeed the farmers and their families, but for whose cooperation and help this study could not have been undertaken and successfully completed. The sample farmers provided unreserved cooperation in allowing us to interview them two-to-three times a week for more than a year and in providing the information without any reluctance. I am grateful to all of them for their splendid and generous treatment of my entire team. They treated me as a member of their own community with a sense of trust and confidence. I salute them for their sincerity and for all the love and understanding they showered on us while we were engaged in the field work.

I am grateful to the field investigators who worked so diligently in collecting data despite all kinds of hardships of rural life to which they were subjected. In particular I wish to record with gratitude the services rendered by the following individual members of the Farming Systems Research Field Team involved in data gathering: Charles Savadogo, Oumarou Kabore, Dandy Quebila, Aime Zongo, Bakary Keita, Lamourdia Kiada, Bara Maro, Larba Bonkoungou, Etienne Dipama, Souleymane Ouminga, Abdou Rasmane, Salifou Buena, Arsene Kabore, Seydou Ouedraogo and Kiri Dianou. Savadogo Kimse (who has since completed his Ph.D. from Purdue and who is now on the faculty of the University of Ouagadougou) did a super job of work as a supervisor and significantly contributed to the success of the project. Thanks are due to Paul Christenson, the agronomist, and Richard Swanson, the anthropologist, on Purdue's Farming Systems Research team, who assisted me in a variety of ways in planning and conducting the field research in Burkina Faso.

While at Purdue, I had the privilege of working with excellent individuals like Kelley White, Woods Thomas, James Colloms, Bill Morris and Earl Kehrberg, who provided me all kinds of support which enabled me to carry out my work as a team leader of the project in Burkina Faso and later to conduct the analysis of data at Purdue after my return from field work. Katy Ibrahim was always there to take care of the administrative problems associated with the West African research project. I am grateful to all of them for their kind support and encouragement.

USAID provided the financial and logistic support to the project, while the Mission Director in Burkina Faso supported the Purdue team by rendering all kinds of services and assistance that made the project a success. I am thankful to them for all their support and help.

Special thanks are due to the government of Burkina Faso (Ministry of Rural Development in particular) and the Organization for African Unity (OAU) representative Akadiri-Soumaila for the guidance, encouragement and assistance they rendered in the conduct of the research project in the region. The comments and suggestions provided by the other researchers working at the national agricultural research station at Kamboinse in Burkina Faso, in particular those of Asnani,

Pattanayak, Matlon and Diemkoumar were extremely helpful to my work in the field and later in the analytical work.

The work on Farming Systems Research--the source of material for this book--was an interdisciplinary work. It involved social scientists, as well as crop scientists that included agronomists, plant breeders, plant pathologists and soil scientists. I took full advantage of my associations with the group of professionals all of whom shared the single goal of raising the living standards of the farm people, the vast majority of whom are poor and are denied the basic amenities of life.

The work of data processing and analysis was done partly at Purdue University and partly at Illinois State University, and I wish to record my sincere thanks to both the institutions for all the facilities they provided me. My graduate students at Illinois State University, Choong S. Tark and Pyeng Bark, were instrumental in doing most of the computer work involving processing, cleaning and analysis of the farm and household data. Bark's Master's thesis, that used my data on crop production and the use of animal traction on small farms was of particular help to me in my work on small farm production systems, and I thank him for all the help and assistance he so willingly provided me. The other students who have helped me include Julie Sykes, Meena Thomas, and Andy K.M. Li, and I thank them all for their assistance and for the work they so diligently performed and thereby contributed toward the successful completion of my work in one way or the other.

I wish to record a deep sense of gratitude to my colleague Mathew J. Morey whose unique expertise in econometrics and programming helped me in my work. Chapter 3, Time Allocation, Home Production and the Economic Contributions of Women and Children, heavily draws upon our joint work and the jointly authored contribution that has been published in the Economic Development and Cultural Change (copyright 1987 by The University of Chicago, 0013-0079/87/3504-0041 $1.00). I have also drawn upon the material reported in a paper jointly authored by myself and Professor R. Ram, and since published in *World Development*. Aneyti Sowa worked with me on a joint working paper on migration using the household data from Burkina Faso, and I sincerely acknowledge his contribution and his assistance in analyzing the survey data in Chapter 4.

Despite their busy schedules, Professors T.W. Schultz, D. Gale Johnson and Gary Becker of the University of Chicago were kind to go through the earlier drafts of some of the chapters and offered critical but extremely valuable comments. I owe to them a great deal for their constructive and thought-provoking evaluation and suggestions. I greatly benefitted from the workshop presentations of my work at the University of Chicago on the economics of farm production and polygyny. In fact, it was during my work as a post-doctoral fellow at the University of Chicago, Department of Economics, that I recognized the importance of the new household economics, human capital, and farming systems and their relevance in developing countries.

I am grateful to Professor Virginia Owen, Dean, College of Arts and Sciences, Illinois State University, for her continuous encouragement and support. I have benefitted from her comments during my several conversations with her on the role of women and children in developing countries. I am grateful to Professor Kenton Machina, chairperson of the Department of Philosophy, Illinois State University, for his valuable assistance in converting the disk-stored manuscript for use on a different word processor.

Sandra Krumtinger typed the draft version of the manuscript and I offer my thankful gratitude to her for the high quality of work she performed and for the tremendous patience she demonstrated.

The final word processing of the manuscript was done by Steven Barrowes of Electronic Secretary. Steven not only did a job of high quality in typing but also helped me in restructuring some of the tables and paragraphs. I benefitted from his intelligent questions and comments. I offer my appreciation for his excellent work.

Kellie Masterson of the Westview Press critically went through the earlier version of the manuscript and offered valuable comments and suggestions, which I very much appreciated.

Lastly, though not least, I owe an unmeasurable amount of debt to my wife, Pravesh, who has supported and encouraged me all along in my efforts to undertake my assignment in West Africa and later to initiate and complete my work on the manuscript; and to my children, Renu and Rahul whose love, affection, and understanding provided me and my wife emotional support that we needed while we lived and worked in Africa, separated from them most of the time.

Ram D. Singh

PART ONE

Introduction

1

Introduction: The Data and the Sample Region

Most of the people in the world are poor, so if we knew the economics of being poor, we would know much of the economics that really matters. Most of the world's poor people earn their living from agriculture, so if we knew the economics of agriculture, we would know much of the economics of being poor....People who are rich find it hard to understand the behavior of poor people. Economists are no exception, for they too find it difficult to comprehend the preferences and scarcity constraints that determine the choices that poor people make.

T.W. Schultz, Nobel Lecture,
Nobel Foundation, Stockholm,
Sweden (December 10, 1979)

I. INTRODUCTION

Africa is a continent of enormous size and great diversity. Although there are some major differences with regard to agro-climatic conditions, natural and human resources, and the geographic sizes across the countries inhabiting the continent, there is also a considerable homogeneity within the region. For example, African economies are for the most part small in economic terms, low average incomes and small populations. The majority of the population in most parts live in the rural areas; 80 to 90 percent of the labor force is employed in agriculture which is predominantly traditional. Most of the African nations suffer from a scarcity of skilled and educated manpower, a heavy dependence on two or three primary export commodities, a low Physical Quality of Life Index (PQLI), widespread poverty, and rapid population growth. Of the thirty countries classified by the United Nations Conference on Trade and Development (UNCTAD) as "the poorest" in the world, twenty countries are located in Africa. Of the thirty-six countries listed as "low income" countries (with a per capita income of less than $370, 1981), almost two-thirds are in Africa. The majority of the poor live in the rural areas and operate small subsistence farms with miserably low levels of crop yields per unit of land. Not only that the agricultural productivity is extremely low but also many of the sub-Saharan regions have frequently faced drought and, consequently, famine conditions. Nine of the twenty-nine sub-Saharan countries (notably, Burkina Faso, Niger, Mauritania, Mozambique, Ghana, Zimbabwe, Mauritius, Angola and Nigeria) experienced negative growth rates in their agriculture, ranging from -0.3 percent to -4.1 percent per annum, during 1970-79 period (The World Bank, 1981). In the 1970s, population growth (of 2.7%) outstripped agricultural production growth (of 1.3%); and twenty-one out of the thirty-nine countries in the region have experienced declining per capita agricultural production (The World Bank, 1981).

However, despite the enormous size of the continent and its obvious importance, economic research for most of its many countries is meager, and often nonexistent in several important areas. The main reasons for this seem to be that most African nations are late comers into the community of independent sovereign states, and partly in the scarcity of satisfactory data even with respect to some of the most basic economic dimensions. The (U.S.) National Academy of Sciences, in its deliberations on African Research Capabilities (1974), rightly comments: "A thorough knowledge of the existing farming systems of Africa is a prerequisite for any kind of research activity in agriculture in Africa that is expected to be put to use....At present no such thorough knowledge exists...and primary data is lacking on farming systems; it must be collected from working farms." Our understanding of the economic conditions of the poor people is indeed very poor. Among other things, the lack of data has prevented: (a) in-depth analyses of the household behavior with respect to major economic decisions that influence family income and welfare, and (b) identification of the constraints that limit household's ability to participate in, and benefit from, economic development. Often, development is said to have bypassed low income households in the low income countries (Lele, 1981). In Africa it has bypassed the subsistence farmers, and in particular the rural women, although the women's role in the African farming system is, unlike that of their counterparts in other developing countries, unique, indeed, "par excellence" for African agriculture (Boserup, 1974).

It is, therefore, of great interest to collect empirical evidence and to critically analyze and interpret the evidence from different settings of the so-called developing nations, especially from those of the sub-Saharan African nations for which primary (household) data is lacking. Such evidence will enable us to comprehend, in the setting of traditional farming systems, the economics of the low income households, the major economic decisions these households make, the basis and the consequences of such decisions, and the conditions and constraints that surround the household's decision making process involving the household and the farm. Equally important, it will help in drawing inferences that may be relevant to research and development, and the role of public policy.

The present effort is an attempt in that direction. The studies reported in this volume provide, with the help of a unique set of data gathered from Burkina Faso (formerly called the Upper Volta), insights into some of the major aspects of the household's economic behavior. This encompasses inquiries into: (a) the household economic decisions--the decisions about marriages, about the number of wives, and about the number of children demanded by households; (b) the time allocation decisions across home production services and farming activities, and the economic value of home production; (c) the household migration decisions, and (d) the farming system of small farms--their production pattern, the estimates of crop production functions, the economic evaluation of the use of modern inputs and animal traction in a predominantly traditional farming system, and the constraints faced by the subsistence farmers.

The study region of Burkina Faso is noteworthy in West Africa not because of any noticeable or remarkable developmental achievement made currently or in the past, but because of a comparatively large international investment in agricultural research financed by various foreign agencies through 8 to 10 research groups located in this region. The major share of such financial aid comes from the United States Agency for International Development (USAID), the United Nations Development Program (UNDP), and the French Development Fund (FAC). Some of the research centers have been working in the region for only a few years, whereas institutes like the Tropical Agricultural Research Institute (IRAT), a French institute, has been in this region for several decades. A considerable amount of foreign assistance has poured into this small landlocked country without any significant impact on agriculture with the exception of cotton production and irrigated rice. A better understanding of the existing production methods practiced by the Voltaic farmers and the economic constraints confronting them may help us appreciate the problems of low productivity and of low farm incomes, indeed, of rural poverty, and aid research that may offer solutions to some of the several complex problems. Furthermore, and importantly, the study will fill an empirical gap in the existing literature on African LDCs. The availability of a well supervised and intensive household survey data set for testing the several hypotheses makes this study unique; perhaps, this is the first one that undertakes indepth and integrated analyses of household economics and the farming system for an African LDC's setting using such a rich (household) data set. Although the data are drawn from Burkina Faso, in view of the historical and structural similarities across countries in much of the erstwhile French West Africa, the study will throw light on several important economic questions for the entire West African region.

However, it also needs to be pointed out that the findings reported in this volume will be subject to the usual limitations associated with cross-sectional analyses; and hence, the need for caution while drawing any strong conclusions and/or making any generalizations based upon the results of the present study.

II. THE DATA AND THE SAMPLE REGIONS

The Data Set

The studies reported in this volume draw upon the data gathered from three sample regions of the central "Mossi" plateau located in the Burkina Faso region of western sub-saharan Africa. Data collection was part of a farming systems research project conducted by Purdue University with financial and logistic support from the United States Agency for International Development (USAID) as a part of a broader regional project on Semi Arid Food-Grain Research and Development (SAFGRAD).[1] Within the Mossi region, a three-part stratification of the central plateau was implemented in order to select: (i) the regions within the country's rural development organizations called the ORDs, (ii) the villages within the ORD region, and (iii) the households within a village. From the three ORD regions, seven villages and 105 households were selected in the initial stratified random sample. However, selection was also based on a desire to ensure representation of important characteristics such as rainfall, soil type, crop patterns, and the relative potential for yield-increasing technology, (Singh 1981, 84). Once the region and village selections were completed, the households were selected randomly from the villages. Only sixty households in five of the seven villages were retained for intensive day-to-day socio-economic observations and agronomic field research.

Prior to the start of the actual field surveys the investigators and supervisors spent a considerable amount of time familiarizing themselves with the people of the sample villages through community meetings and person-to-person contacts. During this period (which lasted for about two months), complete lists of households in each of the sample villages were prepared covering the households' demographic and crop information. This helped in drawing the sample of households. The samples were drawn in the presence of the entire village community. The household data were gathered with the help of pre-tested and structured questionnaires through personal interviews conducted by trained local field investigators and graduate supervisors. The investigating personnel stayed in the villages for the entire survey period (the survey period covered the agricultural year 1980). Each sample household was visited one to two times a week and intensively interviewed during the entire agricultural production year of 1980. This helped in capturing the important seasonal characteristics of home production and farm related activities; carrying the socio-economic surveys across the slack season as well as the peak and semi-peak seasons was considered particularly useful for the study of labor time usage of family members.

The questionnaire was constructed to obtain information on crop production, ranging from soil type, crop rotation and rainfall patterns to the timing of specific operations and input use by crops and by individual fields; on time allocation of family members across approximately one hundred distinct household and farm activities; on the demographic and socio-economic characteristics of the family, such as number of wives, literacy, mortality and fertility; on the family's economic transactions such as buying and selling; and on the value of livestock, capital assets, and the availability of financial credit. The socio-economic surveys were allied closely with the agronomic research that was being conducted concurrently in the sample villages. The farmers in the sample lent unreserved cooperation and support in giving information throughout the entire survey period. In view of the complexity of the surveys and the farmers' farming activities and the time constraints facing them during certain periods, interviews were conducted in household compounds and in the family fields. Furthermore, the investigators were able to gain a first-hand understanding of farm production systems and labor use patterns, as well as the pattern of the household time allocation. Data on yield estimates, demographic characteristics and labor time allocation were particularly checked and verified through on-the-spot observations. The crop yields, for example, were observed in farmer's fields and estimated by actually weighing the crops from groups of lots randomly selected from the fields.

The data base of the project is considered unusually good because of (a) the quality of supervision by highly qualified and experienced supervisors, (b) the intensity with which field surveys were conducted by trained investigators, and (c) the cooperation of farmers in providing the information. The uniqueness of the data set gathered in Burkina Faso arises from its richness in depth and quality. Additionally, household data of this caliber are hard to come by, especially for most of the sub-Saharan (west) African countries. Although the data are from Burkina Faso, because of the historical and structural similarities across the sub-Saharan countries (particularly the erstwhile French-Africa), the characteristics and findings based on these data should be applicable to several countries in the region.

The Sample Regions

The three ORD areas, Ouagadougou, Ouahigouya and Zorgho selected for study in the first phase of the Farming Systems Research during 1979 and 1980, are in the central region of the country. In terms of agricultural potential, the Ouagadougou and Zorgho regions have been categorized as "poor" and the Ouahigouya as "very poor."[2] The three study areas have much higher population density (35 to 43 persons/square km) than the rest of the country (18 persons/square km).[3] The population pressure on cultivable land is, as shown in Table 1.1, the highest in three regions under study.

The Sample Villages

In all, seven villages were selected from the three ORD regions, and their location and population size are presented in Table 1.2. Extremely large or extremely small villages were deliberately excluded from the sample for the purpose of the intensive socio-economic and/or agronomic studies.

TABLE 1.1
Areas and Populations of Three Sample Regions, Burkina Faso

Study Region/ORD	Total Area km^2	Total Population (1000)	Density km^2
Ouagadougou	24,179	847.6	35.1
Zorgho/Koupela	9,039	272.6	30.2
Ouahigouya/Yatenga	12,239	531.5	43.2
National Average	--	--	17.9

Source: Ministry of Planning and Rural Development, 1970-1975.

TABLE 1.2
Households and Population in Sample Villages

ORD Region	Villages	No. of Households (Menages)	Present Population[a]	Persons Living Outside	Total Population
Ouagadougou	Nedogo	208	1,855	263	2,118
Ouahigouya	Sodin	137	983	351	1,334
	Aorema	89	1,020	208	1,228
	Tougou	141	913	284	1,197
Zorgho	Digre and Bissiga	151	1,014	351	1,365
	Tanghin	214	1,809	413	2,222
	Gandaogo	200	1,184	454	1,638

[a]As of the date of Enumeration (1979).

Source: Farming Systems Research, Sample Survey, 1979-80.

The term village[4] is used as per the administrative definition, and hence it generally includes several hamlets (or *quartiers*) scattered around the main or central village where the village chief resides. The hamlets have their own individual names and sub-chiefs. How closely or loosely the hamlets are related to, and integrated with, the main village depends, in most cases, upon the socio-political status and the economic power of the chief. The chief (locally called *Nabha*) is the administrative head of the village who has traditionally provided a vital link between the local administration and the people in his village.

The Sample Households

The households selected in the sample for intensive socio-economic and agronomic surveys provide the data base for most of the studies reported in this volume. It is, therefore, useful to provide a brief descriptive view of these households before presenting the results of the individual studies.

A household in most "Mossi" villages generally consists of the chief or head of the household, his wife, or wives, and their young children. Sometimes married sons and other relatives also stay in the household compound (called concession). Similarly, every household in the village farms some land area and thus there is no class of the so-called landless agricultural workers in rural areas of this West African region as is the case in most other developing countries (of Asia and Latin America).

The data presented in Table 1.3 shows the size distribution of the sample households by age and sex in the 5 sample villages.

On the average, there were slightly over 10 persons present in the household. If, however, one were to include also the absent (migrant) members of the household the average size of the household would be 13 persons. The sample villages in the Ouahigouya region have relatively larger household sizes than the other two regions. The average size of sample households in the sample regions ranges from 8 persons to 12.7 persons (Table 1.4).

The data on migration indicates that 81 percent of all sample households had family members living outside of the region. About 33 percent of the total population lives outside of the villages as out migrants. It is important to note that the migrant members constitute an important part of a household's socio-economic structure in Mossi villages where the percentage of population migrating out side of the region is probably the highest in the country. Such a high rate of migration has been caused by acute pressure of population on land, and lack of alternative economic opportunities in the region. It is useful to provide at this stage an overview of some of the major characteristics of the sample households through the summary data presented in Table 1.5.

TABLE 1.3
Mean Number of Persons Per Sample Household by Age and Sex in the Three Regions, Burkina Faso, 1980[a]

	Oauga Region		Zorgho Region				Ouahigouya Region			
	Nedogo		Digre		Tanghin		Aorema		Sodin	
(Years)	M[b]	F	M	F	M	F	M	F	M	F
0-6	1.5	1.4	1.1	1.6	1.0	1.4	1.8	1.6	1.2	1.8
7-9	1.3	0.4	0.4	0.7	1.0	0.7	0.4	0.6	0.8	0.7
10-14	0.9	1.0	0.4	0.5	0.8	0.3	0.6	0.3	0.9	1.0
15-18	0.5	0.8	0.1	0.4	0.6	0.4	0.7	0.8	0.5	0.7
19-21	0.6	0.6	0.2	0.2	0.3	0.2	0.1	1.0	0.5	0.5
22-55	2.4	2.5	2.1	2.4	1.9	2.0	1.7	2.2	2.5	2.7
56-59	0.2	0.1	--	--	0.1	--	--	0.1	--	0.4
60-64	0.1	0.2	0.2	--	--	0.1	0.2	0.3	0.1	0.1
65-69	0.1	0.1	0.1	--	0.1	--	0.3	--	0.2	--
70+	0.2	0.3	--	0.1	0.2	--	0.4	0.3	0.6	--

[a]Includes all members present and absent.
[b]M = Male and F = Female.

Source: Farming Systems Research, Sample Survey, 1979-80.

TABLE 1.4
Number of Present Members Per Sample Household, 1980

Village (Region)	Number of Persons per Household
Nedogo (Ouagadougou)	11.3
Sodin (Ouahigouya)	12.7
Aorema (Ouahigouya)	11.2
Digre (Zorgho)	8.6
Tanghin (Zorgho)	8.0

Source: Farming Systems Research, Sample Survey, 1979-80.

TABLE 1.5
Some Major Characteristics of Sample Households, Burkina Faso, 1980

	Mean	Standard Deviation
Number of wives per household head	2.37	1.30
Number of others' wives living in household	1.27	1.96
Number of other members in the household	1.29	2.15
Average age of the household head's wives	40.30	14.70
Average age of the household head	56.20	15.00
Number of children born per head of household	11.60	6.60
% of children who died in the age of 0-4 years	32.03	16.04
Average school year of the household head	0.62	2.58
Land area farmed per household (hectares)[a]	4.90	3.25
Total farm revenue in CFA francs[b]	167,233	126,670
Total value of farm capital (in CFA francs)[b]	22,113	27,620
Total value of livestock (in CFA francs)[b]	204,691	544,693

[a]One hectare = 2.47 acres.
[b]One US$ = CFA francs 225-250.

Source: Farming Systems Research, Sample Survey, 1979-80.

A number of interesting phenomena pertaining to rural families in Burkina Faso are indicated. These relate to the polygynous nature of marriages, the high rates of fertility among poor households, the high mortality rates among the small children, the small farm size, the low productivity and low incomes, and the low level of human capital with the average 0.6 years of schooling for the household head (all wives were reported illiterate). However, it needs to be pointed out that the data presented in this section are intended to provide some basic background information to serve as an aid in understanding the discussions of the material contained in the chapters that follow. Hence, any detailed discussion on the individual aspects or characteristics of the data is not warranted at this stage.

The remainder of the book is organized as follows. Chapter 2 is devoted to a discussion of the economics of polygyny and the household demand for children with focus on the major economic factors that seem to influence (a) the marriage decisions--the number of wives the household head acquires over his life cycle and (b) the household's fertility behavior--the number of children the household head plans or desires to produce. Chapter 3 analyzes the household's time allocation decisions and evaluates the economic contributions of farm wives and children to home production, while Chapter 4 discusses the migratory behavior of household members and focuses on the extent to which farm households participate in the external labor market. Chapter 5 presents a discussion on the economics of small farms and traditional farming systems using both descriptive and production function analyses. Chapter 6 is devoted to an economic analysis of the animal traction technology in the setting of an otherwise highly human labor-intensive production system that is still dominant in the agriculture of the region. Chapter 7, the last chapter, highlights the role of agricultural research in the region, the new farm technology and some of the major adoption problems and prospects.

NOTES

1. The other participating institutions were the International Crop Research Institute for Semi-Arid Tropics (ICRISAT, an UNDP project) and the International Institute for Tropical Agriculture (IITA based in Nigeria). The Farming Systems Research was under the Purdue's Farming Systems Unit (FSU).

2. There were in all 11 ORDs (Regional Development Organizations) which are geo-political regions covering the entire country. These are supposedly autonomous organizations responsible for agricultural extension services, credit, marketing, and rural infra-structure. The ORDs in the Western Region (Bobo, Banfora, Diebougou and Dedougou) are characterized as "good" (except Dedougo as "fair"), while in the Eastern region, Fada N'Gourma is characterized as "mediocre" by the Ministry of Planning and Rural Development.

3. The population density for some other regions is as follows: Koudougou, 27.9; Kaya, 27.8; Diebougou, 20.6; Dedougou 16.2; Bobo, 11.5; Banfora, 9.8; and Fada 6.

4. Of the seven villages, only two villages (Nedogo and Digre) were enumerated completely covering the hamlets (quartiers) belonging to the village for administrative purposes, while in the other five villages, some of the hamlets were left out for the sake of convenience.

REFERENCES

Boserup, E. *Women's Role in Economic Development*. London: St. Martin Press, 1974.

Government of Burkina Faso. *Census Reports*. Ouagadougou, Burkina Faso, 1970 to 1975.

Lele, Uma. "Rural Africa: Modernization, Equity, and Long Term Development." Paper #81:31, Department of Economics, The University of Chicago, Chicago, IL, 1981.

The National Academy of Sciences. *African Research Capabilities*. Washington, D.C., 1974.

Schultz, T.W. "The Economics of Being Poor." Nobel Lecture, Stockholm, Sweden (Nobel Foundation), December 1, 1979.

Singh, Ram D. "Small Farm Production Systems in West Africa and Their Relevance to Research and Development." Agricultural Economic Workshop Paper No. 81:20, The University of Chicago, Department of Economics, May 28, 1981.

World Bank. *Accelerated Development in Sub-Saharan Africa: An Agenda for Action*. Washington, D.C.: The World Bank, 1981.

PART TWO

The Economics of the Family

2

The Economics of Polygyny and the Household Demand for Children[1]

I. INTRODUCTION

Recent evidence indicates that polygyny continues to be practiced in most parts of Africa without any significantly noticeable change in its incidence over time; in fact, in some cases, the incidence of polygyny appears to have increased. However, except for the study of Grossbard (1976) and a rather descriptive but interesting survey by Boserup (1974) of some earlier work, the explanations offered for the practice of polygyny in the African setting are mostly in terms of cultural and traditional factors including men's attitudes toward women and the father's desire to perpetuate, through children, his and the family name and prestige (Fuller 1959; Messenger 1959; Schneider 1959; Ohenburg 1959; Wolfe 1959; Lystad 1959; and Dorjahn 1959). Such explanations, although valuable, fail to recognize that social and marital institutions reflect rational decisions about economic welfare, and that there are important economic factors that determine family decisions. Scarcity of cross-sectional household data, among other things, has prevented any detailed economic analysis of household decisions with respect to polygynous marriages and, hence, our understanding of the economic determinants of family marriage decisions in the polygynous societies of Africa remains limited.

The study reported in this section attempts to provide, on the basis of a recent household Survey data set from Burkina Faso, an economic explanation of polygyny and household's fertility behavior in the setting of a highly traditional and predominantly female farming system, and a socio-economic environment that allows farm wives to play a significantly valuable role.[2] The explanation lies in the value of the wife's labor services in farming and non-farming activities, and in the value of children the wife bears and rears. In a setting where an external labor market does not exist, where off-farm (wage paid) employment opportunities within the region are conspicuous by their absence, and where the woman's mobility outside of the village is absolutely restricted, the amount of land farmed by household (male) heads becomes the key determinant of the economic value of the wife's labor services. The demand for the number of wives is essentially a derived demand from (a) the household's demand for wife's own labor services as farm worker, and (b) the household's demand for child quantity which eventually provides another valuable source of labor.

The major objective of the study presented in this chapter is to evaluate the effects of some important farm and family variables on household's marriage and fertility decisions, such as the effects of the size of farms operated and income earned by households, the value of livestock and capital assets, the schooling and age of the husband, and the presence of the other members' wives in the household[3] who may be substituting for the labor services of the household head's wife. The interrelationship between polygyny and the household's demand for children has been particularly focused.

The importance of the study can be seen in several ways. First, it fills an empirical gap in the existing literature by providing evidence on the economic aspect of polygyny especially for a sub-Saharan African LDC's setting in which polygyny continues to be popular and relatively widespread. Second, the labor-value explanation of polygyny attempted in the study fits very well with the socio-economic setting studied because of the widespread participation of rural wives in farming and non-farming activities. Finally, the availability of a well-supervised and intensive household survey data set for testing the economic model of polygyny makes this study unique.

II. THE MODEL AND THE HYPOTHESES ON DETERMINANTS

Following the basic premises of the theory of marriage developed by Gary Becker (1973, 1974, 1974, 1981) as applicable to polygyny,[4] and making the necessary assumptions (such as an uniform production technology, homogeneous productivity of women, and given prices for factors other than own time) a demand function for the number of wives is specified in a semi-reduced form as follows:

$$NCW = a_0 + a_1LND + a_2TRD + a_3TTB + a_4TEV +$$

$$a_5TLV + a_6AAH + a_7YEH + a_8NWO, \quad (1)$$

where NCW = the total number of wives per household head; LND = farm size measured by the amount of land area farmed by the head (i.e., the cropped area in hectares); TRD = farm annual income generated through crop production on the farm (in CFA francs); TTB = total number of children born per head of the household; TEV = value of farm capital assets (in CFA francs); TLV = total value of livestock (in CFA francs); AAH = age of the head of household; YEH = number of years of schooling of the head of household (all the wives in the sample were illiterate); and, NWO = the number of wives of other household members.

In the above formulation of the polygyny model, the following points need to be stressed. First, this model is used to explain men's optimizing behavior in respect of marital output by deciding on the number of wives (i.e., expanding output at the extensive margin) in a socio-economic setting as in Africa that permits polygyny. Second, since the cost of acquiring an additional wife is, by and large, constant across the rural households studied, cost as a specific variable is not included in the model. Third, the wives in this setting contribute significantly to the farm labor supply by their own participation in farming and by producing children who provide the husbands valuable labor services. In this setting all or most labor needs on the farm are generally met by the household; there is hardly any hiring-in of labor from outside of the family, since there is no organized labor market where labor services for farm work can be bought (or sold).

The first major hypothesis that is being examined here is with respect to the land, or the farm size variable. The amount of land operated by the household is considered a most significant variable that determines the productivity of the household head. Differences in farm size are, therefore, assumed to explain the differences in the productivity of husbands across households and thereby the differences in the number of wives that men tend to acquire. In a dominantly agrarian socio-economic setting characterized by traditional, labor-intensive farming system as in Burkina Faso and in other parts of Africa, the amount of land farmed can be a major source of inequality in the productivity of men; and, as Becker points out, "inequality in the various traits among men would be needed to explain polygyny." In rural Africa, additional wife or wives will be more productive even under the condition of diminishing marginal products of men and women within each household. The demand for an additional wife reflects, in a setting such as this, the demand for an additional value of labor. However, this is not to suggest, that women in such societies are treated as mere economic goods to be used (by husbands) as a source of economic gain through their labor services.[5]

The second hypothesis is with respect to the number of children demanded in the household. Since the wife produces children, and the children provide the household head both consumption services and a valuable source of labor, it can reasonably be expected that the demand for the number of wives is positively associated with the number of children, other things constant. Like the farm-size, the underlying point with regard to this hypothesis is that aside from other considerations and variables, polygyny is a function of the demand for the wife's services--in this case for the wife's role as producer of children and, hence, of (children's) labor services.

However, there is an important difference between land, the farm size variable, and the number of children variable. The farm size, i.e., the amount of land farmed can be treated, in the setting studied, as an exogenous variable. The reason for treating land as an exogenous variable is as follows. Good quality land is scarce, and there is no land market where land can be bought and sold. In fact, the same holds for most other rural regions of West Africa. In rural Burkina Faso land is traditionally owned by the village community as a whole and is administered by the tribal chief. Individual families technically do not own the land in the sense that they can not sell (and/or buy) it, or even transfer it to others in the village without the approval of the chief of the village. In some cases, the chief has a council of members chosen by him, or by the villagers, as the case may be, to advise him on administrative and community matters including the distribution of land. In such a communal or tribal land tenure system, the question of the individual titles to land, or of the legal recognition of such a title, or ownership, as we understand it in the western countries, does not exist. Therefore, individuals cannot expand land area by resorting to the market as such a market does not exist. If and when fallow land is available, the village chief allots it to individuals. Such land is generally very poor in quality, located (in the bush) far away from the village. However, the heads of individual household, generally the husbands, cannot be dislodged from their farms under normal circumstances; and after the head's death, his eldest son (or the eldest brother) becomes the head of the household and assumes responsibility for running the household and the farm.

On the other hand, the number of children in the household is truly an endogenous variable. There is a two-way, simultaneous relationship between the number of wives, and the number of children in the household. To resolve this simultaneity problem, another equation for the number of children of the following form has been formulated. This allows for the estimation of the parameters of the model using the two-stage least squares. Equation 2 for the number of children (TTB) can be assumed to represent the household demand for child quantity.

$$TTB = b_0 + b_1LND + b_2TRD + b_3TEV + b_4TLV + b_5AAW + b_6YEH + b_7NOM + b_8NCW + b_9RCM, \quad (2)$$

where all the variables have already been defined except AAW, the average age of the head's wives; RCM, the rate of mortality among children (i.e., % of 0-4 years olds who died); and NOM, the number of other members (relatives) living in the household.

Using Equations 1 and 2, the study tests the hypothesis that the demand for multiple wives, polygyny, is essentially a derived demand from the household demand for: (a) the wives' labor services, and (b) the number of children. It needs to be stressed that the size of the farm (LND) operated by the household heads is assumed to represent, in the framework of this study, the household demand for labor services. Given the two endogenous variables (NCW and TTB), and the number of the predetermined variables that appear in the model, the two equations are identified.

III. THE EMPIRICAL RESULTS

This section has two parts: part one provides a descriptive account of some of the characteristics of polygyny as practiced in the region; while part two, which is the major part of this chapter, presents the results of statistical analysis of the household (polygynous) marriage and fertility decisions.

Polygyny in the Region: A Descriptive View

A summary of the household data on the distribution of married men by age and the number of wives is presented in Table 2.1.

It is evident from this data that the practice of polygyny is quite widespread among the "Mossi" farmers located in the central plateau of Burkina Faso.[6] Of all the rural sample households, 74.6 percent of the household heads were polygynous with over 18 percent of them having 4 to 6 wives on average. For all the households in the sample, the average number of wives per household head was 2.35 (Edna Loose [1979] reports an average of 2.8 wives in one of her sample villages in Senegal). However, this does not imply that all the married men were or are polygynous. On the contrary, only 39.3 percent of all the married men including the heads of household as well as the other married men living in the household were polygynous. Note that Dorjahn (1959) had

TABLE 2.1
Distribution of Household Heads and Other Married Men in the Household by Age and Number of Wives

Age Group (Years)	Total Number of Married Men	Percentage of Married Men with:					
		One Wife	Two Wives	Three Wives	Four Wives	Five Wives	Six Wives
18-35	4	50.0	25.0	25.0	--	--	--
35-45	9	33.3	44.5	11.1	11.1	--	--
45-55	15	26.7	46.7	20.0	6.6	--	--
55-65	11	18.2	45.4	9.1	18.2	--	9.1
65 and above	20	20.0	20.0	30.0	25.0	--	5.0
All Ages, Heads of Households	59	25.4	35.6	20.3	15.3	--	3.4
All Ages, Other Married Men	76	88.2	11.8	--	--	--	--
All Ages, All Married Men	135	60.7	22.2	8.9	6.7	--	1.5

Source: Farming Systems Research, Sample Survey, 1979-80.

reported that only 35 percent of all the married men were polygynous in the sub-Saharan region. In most cases, the Mossi families include married sons and other relatives who live in the same compound (called *concession*). By tradition, after the death of the head of the household, his eldest son and/or the eldest brother inherits the head's wife or wives, along with land, and becomes the head of the household.

A comparison of the data of this study (Table 2.1) with those reported for earlier periods by Dorjahn and other scholars would suggest that polygyny as a marriage institution has not declined in this region of the sub-Saharan west Africa. On the contrary, the present evidence indicates that the incidence of polygyny is a little higher now than what was reported for earlier periods. Furthermore, the data in Table 2.1 also suggests life-cycle pattern in polygyny; as men advance in age, they tend to acquire additional wives. Over two-thirds of the married household heads acquired the second, the third, and even the fourth wife between 35 years and 45 years of their age; on the average, the number of wives for this group ranged between 2.5 and 3 per husband. On the other hand, over 80 percent of the married heads of households in the age-group of 55 years and above were polygynous with 30 percent of them having 4 or more wives, on average. It is clear from the data that most of the married men who start with a single wife end up, during their life-cycle, being polygynous.

Major Determinants of Polygyny: The Results of Statistical Analyses

The mean values and definitions of the variables included in the estimating equation are presented in Table 2.2.

The coefficients of the household demand function for the number of wives were estimated using Equation 1 with the help of both the ordinary-least-squares (OLS) and the two-stage least-squares (2SLS) methods. However, because of the relative advantage of the 2SLS method over the OLS method in providing unbiased estimates of the regression coefficients, the discussion focuses on the results of the 2SLS estimates. The results of the estimate are presented in Table 2.3.

The estimated coefficient on land area (LND), the size of the farm operated by household heads, appears, as expected, positive and statistically significant at the one percent level. As evident from the size of the coefficient for LND (.2035), of all the variables, the land area farmed by husbands appears to be exercising the strongest effect on the number of wives, NCW (Table 2.3). The estimate supports the hypothesis that the

TABLE 2.2
Description, Mean Values and Standard Deviations of Variables of the Model

Variable Description	Mean	Standard Deviation
Number of wives per head (NCW)	2.37	1.30
Number of others' wives (NWO) living in household	1.27	1.96
Number of other members in the household (NOM)	1.29	2.15
Average age of the household head's wives (AAW)	40.30	14.70
Average age of household heads (AAH)	56.20	15.00
Number of children born (TTB) per head of the household	11.6	6.60
Infant mortality i.e., percent of children who died in the age of 0-4 years (RCM)	32.03	16.04
Average school year of the household head (YEH)	0.62	2.58
Land area farmed per household (LND in Hectares)[a]	4.90	3.25
Total farm revenue (TRD, in CFA francs)[b]	167,233	126,670
Total value of farm capital (TEV in CFA francs)[b]	22,113	27,620
Total value of livestock (TLV in CFA francs)[b]	204,691	544,693

[a]One hectare = 2.47 acres.
[b]$1 U.S. = 225-250 CFA francs (1980-81 exchange rates).

Source: Farming Systems Research, Sample Survey, 1979-80.

household heads who operate relatively large land area, and, therefore, require a large amount of labor input, tend to marry more wives, other things constant. This powerful effect of farm size reflects the effect of the wife's labor services, the wife's economic value, on polygynous marriages. Based upon the results of this study, for every 12 acres increase in the size of the farm, the household head tends to acquire an additional wife, other things equal.

TABLE 2.3
Estimated Coefficients for the Number of Wives Model: Rural Households, Burkina Faso

Independent Variables[a]	2SLS (1)	2SLS (2)	OLS (3)
Estimated Number of Children born per Head of Household (TTB)	.1315	.1228	.098
	(2.17)[b]	(1.87)	(4.98)
Age of Head (AAH)	.0039	.0013	.017
	(.27)	(.08)	(1.76)
School-year of the Head (YEH)	.0044	-.0361	.032
	(.07)	(.70)	(.067)
Number of other's wives in the Household (NWO)	.1197	.0423	-.155
	(1.06)	(.45)	(1.58)
Farm Size in hectares (LND)	.2035	--	.209
	(3.07)	--	(3.93)
Annual Farm Income (TRD)	--	.0047	--
	--	(1.65)	--
Value of Capital Assets (TEV)	-.0012	-.0035	-.0008
	(.26)	(.68)	(.17)
Value of Livestock (TLV)	-.005	-.0006	-.0004
	(1.70)	(1.80)	(1.58)
Constant 'a'	-.1125	.2460	-.4800
R^2	.57	.63	.62
F-ratio	8.01	9.83	12.02
N	51	50	59

[a]Dependent Variable = NCW, Number of Wives per Household Head.
[b]Figures in parentheses are t-values.

The plausibility of the land-polygyny connection must be viewed in the light of some of the peculiar characteristics of the African setting. First, the women provide 62 to 65 percent of the total labor input usage on the family farms in the region studied.[7] Furthermore, wives play a key role in their family's trade and commerce, in addition to their major share in contribution to home production. The estimated dollar value of wives' home production (that comprises of her work in household services such as child care, cooking, cleaning, home maintenance work, collection of firewood and drinking water for the family and production transformation) ranges from $307 to $890 annually. Second, women seem to work harder and make greater contribution to farm production than the men do as evident from the fact that the estimated marginal productivity of the family female labor on the farm is significantly higher than the male labor.[8] Third, families with relatively greater number of wives per head of household have been observed to be practicing a more diversified cropping pattern[9] that allows them to grow cash crops which contributes to the family's cash income. Indeed the role of African women both in the farming system and in the home production is par excellence, and this role provides an important explanation of polygynous marriages. Although, this may not be a complete explanation.

One might wonder whether the causation behind the positive and strong relation between the amount of land and the number of wives is from wives to land rather than from land to wives. The question really is whether land is, as treated in this study, an exogenous variable, or it is an endogenous variable. In the rural setting of Burkina Faso, as in most other African countries, there is generally no land market through which individuals can buy or sell land. As pointed out earlier, there is no individual ownership of, or title to, land as it is traditionally held (owned) by the tribal and/or the village chiefs, although individual families farm the fields and make their own decisions with respect to production and marketing of their produce. Individuals in this setting cannot normally increase the amount of land simply by acquiring more (or additional) wives. Hence, it was reasonable to treat land as an exogenous variable and, further, to expect the causation from land to wives rather than the reverse; and the latter expectation is confirmed by the statistical results of this study.

The study examines another important question, i.e., whether the powerful farm-size effect was also capturing the income-effect, and whether the income variable itself exercised any significant effect on the demand for wives. To evaluate this aspect of the relationship, the polygyny model incorporated family farm income (TRD)[10] as another regressor. Income included in the model was the estimated family's annual net farm income (which equals gross revenue from all crops minus the purchased inputs). Information about other cash incomes, which are in any case insignificant, was not available, and therefore, such incomes were not included. This may have resulted in causing some measurement error in the income variable. Another problem with this variable is that farm income will also contain certain transitory components subject to fluctuations, and that may further weaken the variable. However, the equation for the number of wives (NCW) including the income variable (TRD) was estimated. This equation had all of the variables from Equation 1, except for the land variable, which was not included because of the high intercorrelation between this variable and income.

The coefficient of family farm income, TRD, appears small and statistically not significant (Equation 2, Table 2.3). The size and the level of significance of the coefficient of this variable remained stable and consistent across several equations tried with alternate specificational variations. In fact, apart from the measurement problems mentioned above, the income variable is not a truly exogenous variable, as the family farm income is significantly determined by the number of wives and their labor input. This being the case, one would suspect that the estimated coefficient of income was biased upward. This implies that the income variable has probably an even weaker effect than is indicated by its coefficient and, further, that the farm size effect is not capturing in any major way the income effect.

In a dominantly traditional agrarian society, such as in the region studied , which permits polygyny, and in which the (i) human labor input plays a most critical role in farm production, and (ii) the scope for the economic participation of wives tends to increase with increasing farm size, it is not at all surprising that the farm size shows such a strong, indeed the strongest, effect on polygyny.

The coefficient of the predicted number of children per household head (TTB), appears positive and statistically significant at the 5 percent level (Table 2.3). The positive association between the number of wives and the number of children was expected. The wives produce children and the children provide parents not only consumption services but also production services through their participation in the household's economic activities. The children in the Mossi farm families begin to participate in economic activities from an early age of 7 to 8 years, and thus provide parents valuable labor services of all sorts (such as herding of cattle, farm work, fetching firewood and drinking water). For example, in the sample region studied, children 7-14 years old contribute over 16 percent of the household's total available labor supply. In fact, the children's share in family income and welfare will be much greater if we also add the contribution of the older children. And, since the children through their labor services enhance the economic value of the mother, it is only reasonable to expect that an increase in the number of children demanded by the husband will tend to raise the demand for multiple wives, other things constant.

The coefficient of the livestock variable (TLV) is negative, and statistically significant at the 10 percent level. The negative connection between the number of wives and the value of livestock owned by households seems quite plausible in the African setting. Livestock represents both a sellable and a buyable asset in the household, and the husbands seem to be selling some part of their livestock so as to be able to meet the cost of all sorts of acquiring additional wives, or giving away to the bride's parents animals as a part of the price (or gift) as per the traditional customs in the region. The fact is that of all the assets held by the farm household, livestock is the one, perhaps the only one, which is frequently transacted in the market as and when needed. Land, for example, cannot be bought or sold as per the traditional rules still in vogue in the region. Land is owned by the community and individual households cannot transact it, nor can they transfer it to anyone else outside of their families.

The coefficient of the number of wives of the other members (NWO) appears statistically weak (Table 2.3). The coefficient of the husband's schooling (YEH), although positive, is statistically not different from zero. The polygyny-schooling relationship needs to be considered in view of the fact that schooling among the farmers in Burkina Faso is extremely low with only a 6 to 7 percent literacy rate in rural areas. The household heads studied have had a little over half a year (0.6) of schooling, most of which is below primary school level. Husband's age (AAH) does not show any significant effect on the number of wives. In fact, the farm size-effect seems to be overshadowing the age-effect in explaining the husband's decisions to acquire additional wife or wives.

Household's Fertility Decisions: The Demand for Children

It is recognized that there is a simultaneous relationship between polygyny and the number of children desired by parents, by husbands in general in the setting under study. The results of this study show that a strong association exists between the two. However, the positive association between the number of wives and the number of children needs to be interpreted with a qualification. This is that men do not tend to marry several wives because they expect that each additional wife adds more children than the previous wife does. The decision to marry several wives is related with the household demand for children per husband, i.e., the total number of children desired in the household from all wives.

Given the generally poor health of rural women, the harsh working conditions, poverty, and the lack of basic requirements such as safe drinking water, health care and medical services, it is not surprising that the rate of mortality among rural children in the region is very high, 32 to 33 percent of all children born die between the ages of 0 and 4 years! In a situation such as this, the reproductive capacity of rural women tends to deteriorate fast as they advance in age. One way to overcome this problem for the husband is to marry several wives, because as the number of wives increases, the chances for producing more children for the household head, the husband, may also tend to increase. Household heads appear to be maximizing their objective function for the total number of children rather than for the number of children from each individual wife (i.e., the number of children per wife), that they tend to marry over their life cycle.

The results of the estimates of the fertility equations with the two alternate specifications of the dependent variable, one with the total number of children born to all wives (i.e., the number of children per head of the household), and the other with the number of children born per wife, are presented in Table 2.4. Two important results emerged from the estimates with respect to the polygyny-fertility connection. First, as indicated by the estimated coefficient for the number of wives variable (NCW) in Equation 1, there is a positive and statistically significant relationship between the predicted number of wives (NCW) and the number of children born per head of the household. The result implies that with each additional wife that the husband marries, there is an increase in the total number of children by 3.1 in the household, all other variables

TABLE 2.4
Estimated Coefficients for the Fertility Model: Rural Households, Burkina Faso, 1980

Independent Variables	2SLS (1)[a]	2SLS (2)[b]
NCW	3.131	-1.430
	(2.95)[c]	(12.86)
NOM	-0.360	-0.230
	(-1.90)	(-2.54)
RCM	0.07	0.022
	(1.95)	(1.32)
AAH	0.0617	0.050
	(1.273)	(2.21)
YEH	0.303	0.080
	(0.247)	(0.72)
LND	0.087	0.300
	(0.22)	(1.59)
TEV	-0.0016	0.00012
	(-0.068)	(0.11)
TLV	0.00284	0.00066
	(1.87)	(0.93)
Constant 'a'	-2.320	3.66
	(-0.85)	(2.85)
R^2 (based on instruments)	0.58	0.25
F	8.60	8.91
N	59	58

[a]The dependent variable here is TTB = the number of children born per head of household.
[b]The dependent variable here is TB = the number of children born per wife.
[c]Figures in parentheses are 't' values.

held constant. Second, the coefficient for the number of wives variable in Equation 2 with the number of children per wife as the dependent variable, appears negative and statistically highly significant. This result provides a strong support for the hypothesis that the number of children born per wife tends to decline as the number of wives increases in the household. Thus, although the total (overall) fertility rises the per wife fertility declines, as the number of wives increases in the household. However, the important result to note is that the household's fertility and marriage decisions seem to be influenced by consideration of the total number of children, rather than the number of children per wife, desired by parents. Husbands, the heads of household in polygynous families, are the major decision makers in the process.

The coefficient of child mortality, RCM, appears , as expected, positive (Equation 1, Table 2.4) and statistically significant at the 5 percent level. This result suggests for a positive association between fertility and mortality among children. The mortality rate of rural children is considerably high in this region; as per the estimate made in this study, 32 percent of the children born to married couples die before reaching four years of age. Interestingly, the rate of mortality is relatively lower for the female children (29.9%) than the male children (34.6%); although for most other (non-African) developing countries the data on child mortality shows relatively higher mortality among the female children.

However, the important thing that needs to be focused in this context is that there is a significantly positive connection between mortality and fertility. This would indicate that there exists a dilemma: the fact of a lower survival rate for rural children leads to efforts made by parents to produce more children (i.e., higher fertility rate). The result of such a relationship as found in this study supports the partial replacement hypothesis. Thus, with greater uncertainty about the survival of children, the parents may tend to increase the number of children; this tendency is likely to become stronger when the parents' goal is to have a given number of children. From the view point of an over-populated poor area this may be unfortunate.

The other variable that seems to have some impact of significance on the number of children is the value of live-stock in the household. The effect of this variable is positive; the households with greater number of live-stock also tend to have relatively greater number of children. An important economic activity of children in the household is herding and caring of livestock; in fact, 50 percent of the children's total labor time allocated to household and farming activities is spent on animal care. It is, therefore, quite plausible that the livestock variable shows a rather important effect on the household demand for child quantity.

IV. SOME CONCLUDING COMMENTS

The analysis of the data from the sub-Saharan region yields empirical support for an economic interpretation of the polygynous marriage behavior and the demand for children of the traditional rural farm households. The household heads are utility maximizers, and their marriage and fertility decisions are made so as to maximize the returns from the wives and their children. The factors, the amount of land farmed, and the predicted demand for the number of children influence the observed polygynous marriages positively and significantly. The income effect appears weak and statistically not significant. Both land and children variables represent the economic value of wives for husbands.

The results with regard to the farm size and the predicted number of children indicate complementarity between wives as workers and wives as producers of children. Given the major contribution of women and the importance of children as producers as well as consumers, it is not surprising that a system developed whereby wives provided husbands both labor and children.[11] Additionally, the cost of producing and raising children in the rural areas studied is also relatively small since wives can care for children while working in the household or on the farm. Moreover, since most of the rural children do not attend school, the cost of the children with regard to the quality (schooling) is also minimal. There is, therefore, greater incentive to produce children, and, as a result to acquire additional wives. The estimated coefficients of the land area farmed (LND) and the number of children (TTB) reinforce our contention about the complementarity of wives and its significance in polygynous marriages.

Husband's age and schooling show statistically insignificant effect on the number of wives; it appears, the effects of the land and the child quantity variables overshadowed the effects of the age and schooling variables. The negative association between the number of wives and the value of livestock in the household seems to imply some form of substitutability between the number of wives and livestock.

The results of the fertility model suggest for a strongly positive association between the number of wives and the (total) number of children, while a powerfully negative association between the number of wives and the number of children born per wife. The livestock variable has a positive effect on the number of children demanded by rural parents. This implies that the economic value of children to households is a determinant of child quantity. An important result of the fertility model is the emergence of a positive and statistically significant relationship between mortality and fertility. This evidence appears quite relevant to public policy. If reduction in the fertility rates is a goal of public policy, one way to achieve this goal in the long-run may be through measures which help reduce mortality among young children. Important among such measures could be the provision of health care and medical facilities, information and extension services for health and nutritional guidance. Also, improved sanitation and other forms of assistance under the general health improvement programs of governmental and other agencies could be supplied. Making these facilities accessible to poor households could form a broadly based approach to fertility reduction.

Finally, the results of the model also suggest the usefulness and feasibility of the simultaneous analysis of family decisions, the decision about the number of wives and the decision about the number of children.

The results of the study need to be interpreted in the light of the following limitations, however. First is with regard to the data base of the study. As stated earlier, the polygynous marriage decisions have been explained in the particular socio-economic setting of the region(s) in which the economic contribution of wives to family welfare is quite substantial, and, equally important, there are generally no social or family taboos on the wife's labor force participation. These conditions may not hold for all the polygynous societies, for example, the North African and the Middle Eastern countries. Second, the estimated coefficients, based as these are on cross sectional household data, are not intended to be used for prediction purposes. To attempt to do that we will definitely need a much broader data base across countries and across time, and a model that not only incorporates economic variables but also some of the relevant legal, social and cultural variables. Furthermore, the effects of modern farm technology and education on the practice of polygynous marriages in rural Africa would need to be studied. However, this study was not intended to answer these questions.

NOTES

1. The major part of the material in this section is drawn from the author's paper presented at the Agricultural Economic Workshop, The University of Chicago, Department of Economics, April 19, 1984, (Paper No. 84:12). The thoughtful comments and suggestions offered by Professors Gary Becker (who, in fact, initiated the author to undertake this study), T.W. Schultz and D. Gale Johnson who asked searching questions during the workshop are gratefully acknowledged.

2. In other settings where polygyny is socially acceptable, where men do most of the agricultural work, and where the women's economic role is limited to home production, the incidence of polygyny is substantially small. As observed by Boserup (1974), "the proportion of polygamic marriages is reported to be below 4 percent in Egypt, 2 percent in Algeria, 3 percent in Pakistan and Indonesia. Polygamy offers fewer incentives in these parts of the world where shifting cultivation has been replaced by the permanent cultivation (because of dense population), and where wives do not work as farm workers."

3. As customary, the other members and their wives live in the house compound of the head, and, like any other member, contribute their labor services to the family farm. In the polygyny model (1), only the wives of the other members (NWO) appears, while in the fertility model, all the other members (NOM) appears on the right hand side.

4. The model developed by Becker assumes that each individual tries to maximize his or her utility from marriage (which is voluntary), and further that the market is competitive and is in equilibrium. Gains from marriage are related to compatibility and complementarity of individual's time, goods and other inputs in household production. For details of the model see Becker (1973, 1974, 1974, and 1981).

5. From personal observations and contacts with the sample households during the field surveys in the study regions, Singh (1981) observed that "women in Mossi families are caring and loving social beings as household members as one could expect their counterparts to be in monogamous societies." No cases of maltreatment of wives, or even of family feuds in polygynous families, were reported from any of the sample villages in Burkina Faso during the survey period (1979-81).

6. Of the country's 6.3 million people (1981), 89 percent live in the rural areas and depend on agriculture as the main source of livelihood. The country's census data show a rather

favorable sex-ratio for males with 1.3 females per male population (1975) in the age group of 15 years and about for females and 20 years and above for males. For rural areas, however, the number of females per male is higher, for example, in the sample region the number of females per male population in the above age groups was 1.8. The sample data do not show any difference in the practice of polygyny among the three religious groups (animists, 75%; Muslims, 22%; and Christians, 3% of the total population).

7. As per the estimates made on the basis of household members' time allocation on an average, the wife allocates 4.8 hours per day as compared to 3 hours of the husband, on activities such as farming, animal care, food processing and household activities (cooking, cleaning, child care etc.).

8. See Chapter 5 on the economics of small farms, in particular tables that contain the estimated coefficients for the female labor and the male labor inputs. The marginal productivity of female labor in agricultural production is substantially higher than the male labor.

9. See Chapter 5 for more on cropping pattern and the number of wives of household heads.

10. Family income, TRD, is expressed in '000 CFA francs, and 225 to 250 CFA francs = $1 U.S. (1980).

11. I am grateful to Professor Gary S. Becker who forcefully brought to my attention this point while commenting on an earlier draft of my paper on the subject.

REFERENCES

Becker, G.S. "A Theory of Marriage: Part I." *Journal of Political Economy*, 81(1973):813-46.

__________. "A Theory of Marriage: Part II." *Journal of Political Economy*, 82(1974):S11-S26.

__________. "A Theory of Social Interactions." *Journal of Political Economy*, 82(1974):1063-93.

__________. *A Treatise on the Family*. Cambridge: Harvard University Press, 1981.

Boserup, E. *Woman's Role in Economic Development*. London: St. Martin Press, 1974.

Dorjahn, Vernon R. "The Factor of Polygyny in African Demography." In *Continuity and Change in African Culture*, ed. by W.R. Bascom and M.J. Herskovits. Chicago: The University of Chicago Press, 1959.

Fuller, C.W. "Ethnohistory in the Study of Culture Change in Southeast Africa." In *Continuity and Change in African Cultures*, ed. by W.R. Bascom and M.J. Herskovits. Chicago: The University of Chicago Press, 1959.

Grossbard, Amyra. "An Economic Analysis of Polygyny: The Case of Maiduguri." *Current Anthropology*, 17(1976): 701-707.

Loose, Edna E. "Women in Rural Senegal: Some Economic and Social Observations." Workshop Paper, Sahelian Agricultural Workshop, Department of Agricultural Economics, Purdue University, Feb. 1979.

Lystad, R.A. "Marriage and Kinship Among the Ashanti and Agni: A Study of Differential Acculturation." In *Continuity and Change in African Culture*, ed. by W.R. Bascom and M.J. Herskovits. Chicago: The University of Chicago Press, 1959.

McSweeny, Breda Gael and M. Freedman. "Lack of Time as an Obstacle to Women's Education: The Case of Upper Volta." In *Women's Education in the Third World: Comparative Perspectives*, ed. by Gail F. Kelly and Carolyn M. Elliott, pp. 88-103. Albany: State University of New York Press, 1982.

Messenger, J.C. "Religious Acculturation Among the Anang Ibibio." In *Continuity and Change in African Cultures*, ed. by W.R. Bascom and M.J. Herskovits. Chicago: The University of Chicago Press, 1959.

Morey, M.J. and R.D. Singh. "The Value of Work at Home: Contributions of Wives' Household Service in a Developing Country." Working Paper #8419, Department of Economics, Illinois State University, Normal, Illinois, October, 1984.

Ohenburg, P.V. "The Changing Economic Position of Women Among the Afikpo Ibo." In *Continuity and Change in African Cultures*, ed. by W.R. Bascom and M.J. Herskovits. Chicago: The University of Chicago Press, 1959.

Ram, Rati and Singh, R.D. "Farm Households in Rural Burkina Faso: Some Evidence on Allocative and Direct Returns to Schooling and Male-Female Productivity Differentials." Forthcoming, World Development (1988).

Schneider, H.D. "Pakot Resistance to Change." In *Continuity and Change in African Cultures*, ed. by W.R. Bascom and M.J. Herskovits. Chicago: The University of Chicago Press, 1959.

Schultz, T.W., ed. "Fertility and Economic Values." *Economics of the Family: Marriage, Children and Human Capital*. Chicago: The University of Chicago Press, 1975.

Singh, R.D. Small Farm Production Systems in West Africa and their Relevance to Research and Development: Lessons from Upper Volta. Workshop Paper #81:20, Department of Economics, The University of Chicago, May, 1981.

Wolfe, A.W. "The Dynamics of the Ngomba Segmentary System." In *Continuity and Changes in African Cultures*, ed. by W.R. Bascom and M.J. Herskovits. Chicago: The University of Chicago Press, 1959.

3

Time Allocation, Home Production and the Economic Contributions of Women and Children[1]

I. INTRODUCTION

The economic role of women in home production has been widely recognized by social scientists, notably economists, anthropologists, and sociologists, as well as by women leaders in both the developed and the developing countries. Note, for instance, that the UN World Conference held recently in Nairobi marking the close of the Decade for the Women has recommended, among other things, that the services rendered by women in the household, in home production, be included in the national income accounts so that a country's estimates of gross domestic product (GDP) do not exclude the value of such services.[2] In low income countries where production in the household and production for the market are highly labor intensive, the value of work-at-home often represents a significant addition to family income and welfare, in fact, as pointed out by T.W. Schultz (1974), much more significant than that in the high income countries.

Similarly, it is recognized that children in low countries provide another important source of economic value to households. Children make substantial contributions to their parent's welfare through the work they perform in the household and on the farm, and through the food and shelter they provide for their parents when the parents become elderly. To quote T. W. Schultz (1974), "children are, in a very important sense, the poor man's capital." Parental investment in children, as a form of capital, involves parent's resource allocation decisions regarding the number and quality of children desired. The economic contribution that children make to their family's income stream influences these parental decisions. Although the nature of work done by children varies with the age and gender of the child, the work performed by children often represents a substitution of their time for that of their parents. Studies from India and a few other developing countries provide empirical evidence on the economic contribution of children through the labor time of children spent on farm activities (Cain 1977; Evenson 1970, 1978; Evenson and Popkin 1978; Makhija 1976; Nag, White and Creighton 1978; Rosenzweig 1980; Shortlidge 1976; and Singh 1978).

Often children in developing countries start participating in a household's economic activities from an early age of seven to nine years and by most western standards these children may be considered to be abused. However, in the traditional agrarian socio-economic environment of the developing countries, child labor is accepted as a normal phenomenon. While governments, through legislation, have banned child labor in organized industrial sectors, the use of child labor in home production and on the farm goes unnoticed and unchallenged.

Despite the significance of home production, interest in household economics in developing countries is of relatively recent origin. Most of the studies, reported from the few developing countries, have focused on the time allocation, fertility and schooling decisions of households using, as a basis, the Becker-Lewis model (Becker 1965, 1975; Becker and Lewis 1973). There is hardly any empirical study with respect to the developing countries, the LDCs, that provides estimates of the economic contribution through home production, that is, the "dollar" value of work-at-home. For the developed countries, however, there are a number of studies that have estimated the value of home production using a wide range of statistical techniques (Hawrylyshyn 1976). Gronau's (1973, 1976, 1980) pioneering work in this area provides one of the most recent attempts to estimate home production in a developed country using econometric methods. However, Gronau's estimates are focused on the labor supply function and the marginal productivity of women's work in home production.

The paucity of data from developing countries on the household's time use, or home production, has constrained attempts to estimate either home production functions or the value of home production. Although, in the African setting, the economic role of women and children is significant and in several respects unique when compared to the role of their counterparts in other developing countries (Boserup 1970). The fact is that very little is known about the African developing countries, in particular, in terms of quantitative estimates of the household's resource allocation decisions, much less any quantitative estimates of the value of home production or of the economic contributions that farm wives and other make to the household.

The uniqueness of the African setting stems from several characteristics: the prevalence of polygyny with an extended family system and large households, the highly traditional, labor-intensive, agricultural systems, and the much greater participation of women and children observed in farming activities in this setting than that of the other developing countries. Above all, the households in this setting play a key role in the production of goods, many of which are neither marketed nor directly measurable. But, at the same time, the production of such home-produced goods and services significantly adds to family income and welfare.

It is against the background of the setting described above that the study of home production and the economic contribution of women and children reported in this chapter assumes significance. The major focus of the study is on: (a) the estimation of the household's labor time allocation through (i) a farm labor supply function and (ii) a work-at-home function; and (b) the evaluation of the economic contributions, or the dollar value of work-at-home of rural women and children. Note, that in the national income accounts, the value of home-produced services is not included, although a substantial portion of goods and services produced and consumed by households are home produced. Since national income estimates exclude home production, the (dollar) income data reported for countries would not reflect the true living standards or the economic welfare of people. For countries, such as the African countries, in which home production constitutes a major portion of total production and consumption and in which the women and children contribute a much greater share than their counterparts do in most other developing and/or developed countries, the estimation of a home-production function and of the value of work-at-home indeed becomes meaningful. This is so because it offers some possible economic explanation of the home-production function and quantifies the economic contribution made by women and children to family income and welfare. Furthermore, in the otherwise overwhelmingly traditional agricultural system that has limited the involvement of farm households with the wider economy through the market, the estimation of a home-production function and of the value of work-at-home may also provide a better measure and understanding of the growth process.

The remainder of the chapter is organized as follows. In Section II, an overall descriptive view of the labor time allocation of household members focuses on the major types of activities and the relative shares of household members in the performance of these activities. A brief discussion of the farm labor supply estimated through a time allocation model is also presented in this section. Section III contains the results of the estimated home production function as well as the dollar value of work-at-home of wives; while Section IV discusses the economic contribution and the value of work-at-home of children. The major thrust of this chapter is on the latter two sections, however. The model and the forms of the estimating econometric equations are provided in the appendix.

II. HOUSEHOLD LABOR TIME ALLOCATION AND THE FARM LABOR SUPPLY FUNCTION

Labor Supply and the Relative Shares of Household Members

The summary statistics on labor time allocation presented in Tables 3.1 and 3.2 provide important insights into the pattern of the distribution of time use and the relative shares of the household members' time in home production and farming activities.

Home-production activities in this study include child care, product transformation (milling and cleaning and grinding of grains, e.g.), cooking, cleaning and other maintenance work at home, fetching drinking water, and collecting firewood (for fuel). Activities related to farming (such as land preparation, planting, weeding, harvesting, threshing, animal care, etc.) are not included in home production. All of the activities performed in home production are labor intensive and most of these are performed by the wives, the female children also sharing a part of the household work. There is, in this setting of a highly patriarchal family and social structure, a distinct, largely traditional division of labor between the male and the female members with the female members performing most of the home-production related tasks. For example, on the average, the wife spends about 1.70 hours per day on child care, cooking, cleaning, fetching, water, etc., while the husband spends no more than 0.02 hours on these activities. Additionally, the wife spends on the average, 1.50 hours per day on other home production activities of product transformation. In respect to other activities, there are also distinct sex differentials in the allocation of the time of the household members. Activities such as animal care, fishing, hunting, and house construction are mostly the domain of the male members.

TABLE 3.1
Wives' and Children's Shares in Home Production and Farm Activities: Percentage of Labor Time Use, Rural Households, Burkina Faso, 1980

	Percent of Total Labor Time					
	Children (7-14 Years)		Children Over 14 Years		Hus-	
Activity Type	Male	Female	Male	Female	bands	Wives
Household Type I (product trans- formation, trade/ marketing)	3.0	11.5	19.2	22.4	18.4	22.5
Household Type II (arts, crafts, con- struction, fishing and hunting)	9.8	4.3	25.5	4.9	41.9	13.6
Household Type III (child care, cooking, cleaning, fetching water and collect- ing firewood)	11.3	15.5	4.4	23.8	0.5	44.5
Farming Activities[a] (land preparation, planting, weeding, harvesting and animal care)	8.3	6.9	16.0	15.9	13.0	13.7
Only Animal Care[b]	49.1 (all children)				8.3	5.6

[a]Of the total farm labor time 26.2% was contributed by "other members" (12% by males and 14.2% by females).
[b]Of the total animal care labor time, 37% was contributed by "other members" (19.8% by males and 17.2% by females).

Source: Farming Systems Research, Sample Survey, 1979-80.

TABLE 3.2
Household Members' Mean Hours of Work Per Day Across Major Household and Farming Activities, Rural Households, Burkina Faso, 1980

Activity Type	Husbands	Wives	Children 7-14 Years		Children 15 Years & Above		Other Household Members	
			Male	Female	Male	Female	Male	Female
Household Type I (product transformation trade/marketing)	.83 (1.17)	1.15 (1.27)	.13 (.49)	.52 (.92)	.87 (2.24)	1.01 (1.19)	1.15 (2.17)	.71 (.90)
Household Type II (arts, crafts, construction, fishing and hunting)	1.20 (1.13)	.39 (1.05)	.28 (.75)	.12 (.31)	.73 (1.50)	.14 (.39)	.81 (1.99)	.31 (.94)
Household Type III (child care, cooking, cleaning, fetching water and collecting firewood)	.02 (.13)	1.61 (1.44)	.41 (1.10)	.56 (2.89)	.16 (.42)	.86 (1.02)	.31 (.71)	1.57 (1.93)
Farming Activity (land preparation, planting, weeding and harvesting; and animal care)	3.60 (2.30)	3.80 (2.00)	2.30 (3.20)	1.91 (2.85)	4.43 (2.83)	4.42 (3.09)	3.32 (2.95)	3.95 (2.60)
N (number of cases)	60	113	75	69	60	23	15	58

[a]Standard deviations of means in parentheses.

Source: Farming Systems Research, Sample Survey, 1979-80.

Farm work is shared by husbands, wives and children. In fact, the average number of hours allocated to farm work is slightly greater for wives (3.8 hours per day) than for the husbands (3.6 hours). Also, note that most of the animal-care work is performed by children, the male children providing the major share of labor for animal-care activities.

Another important point to note in this regard is that several of the home-production activities, in particular, cooking, cleaning, and product transformation, can be, and often are, jointly performed along with child care. However, to avoid double counting, maximum care was taken to separately estimate the time use on each of the single tasks while gathering the time allocation data by discussing with the household members their labor-use pattern across several activities on a daily basis.

For all three classifications of home production activities, the children's share appears to be substantial. Furthermore, as the children grow older, their participation in home production as well as in farming tends to increase. As shown by the data, it increases from 3 percent for the 7-14 years old male children to 19.2 percent for the adult 15 years and older male children for Type I home production activities; from 9.8 percent to 25.5 percent for the Type II home production activities; and from 8.3 percent to 16.0 percent for the farming activities (Table 3.1). The distribution of time use classified by gender and age would imply that the services rendered by children in the home and on the farm represent a substitution of the children's time for the time of the parents. This is true especially in home production activities. This leads to the hypothesis that the opportunity cost of the children's time spent on activities other than home production and farming is likely to be greater in this setting than for the children in the developed countries. In a setting where purchased goods are costly and often beyond the means of poor parents, households tend to substitute labor time-intensive home produced goods for market goods. The labor time provided by either women or children thus constitutes a valuable resource available to low income households.

The Farm Labor Supply Function

The variables of the labor supply model and their descriptions are provided in Table 3.3.

The results of the estimated coefficient of the farm labor supply model are presented in Tables 3.4 and 3.5 using the ordinary least-squares (OLS) and the two-stage least-squares (2-SLS), respectively. Focusing first on the household head's

TABLE 3.3
Description of Variables of the Estimating Equations in Tables 3.4 and 3.5

Variable	Description
TFH	Total hours of farm work of husbands
TFW	Total hours of farm work of wives
TFC	Total hours of farm work of Children
TFO	Total hours of farm work of others
CO4	Number of children 0 to 4 years of age
AYC	Age of the youngest child
MC714	Number of male children 7-14 years of age
FC714	Number of female children 7-14 years of age
HAGE	Age of the household head
HED	Years of Schooling of the household head
OED	Years of Schooling of the other members
AREA	Land area farmed in hectares
AT	Animal traction hours
PFI	Expenses (in CFA francs) on purchased farm inputs
NW-1	Number of wives minus one
AW	Average age of wives
EXLTH	Exogenously determined leisure time (on social ceremonies, for example)

labor supply Equations 1 and 2, it is evident that the effect of the wives' total farm time, the variable TFW, is positive but statistically insignificant. In fact, the sign on each of the "time" variables, the total wives' farm time (TFW), and the other members' farm time (TFO) is positive and only the coefficient on TFO is significant (at the 6 percent level). This may imply that little substitution is taking place within these household members. However, the positive association observed for the head's equation (as well as for the wives' equations,) is due to the fact that the timing of farm activity is exogenous, dictated by weather conditions and the natural length of the growing season. Consequently, when the opportunity or need to complete certain tasks at certain critical times arises every working member in the household is involved in farm activity.

There may be some substitution between wives and heads of household as indicated by the significant negative coefficient on the number of wives of household heads (NW) in the second equation. Taken literally, in the average household, for each additional wife, the household head works about 81 hours less during the crop season, ceteris paribus. This may be occurring for two reasons. First, there is substitution taking place in farming activities between heads and wives. Second, the number of wives variable (NW) may also be capturing some age effects since the acquisition of additional wives takes time; there is a positive correlation between the number of wives (NW) and the age of the household head.

Some indication of the economic contribution of children is provided by the coefficient on aggregate time of adult children, measured by the variable, TFC, in both labor supply equations. Both coefficients are negative and in Equation 2 significant at less than the five percent level. The signs on the coefficient of the adult children's total farm time, TFC, (Table 3.4) are also consistent with the signs obtained on the children's total farm time (TFC) in the 2SLS estimation of a system of farm labor supply functions (Table 3.5). Another indication of the economic contribution of children to the household can be seen in the signs and magnitudes of the coefficients on number of male and female children between the ages of seven and fourteen (MC714 and FC714). It was expected that "young" adults (children) would be able to contribute time to certain household activities and to some child care, thus permitting both the head and the wives to devote more time to farm, market and other household activities. This seems to be almost uniformly supported by the results of the single equation estimations (Table 3.4).

A possible proxy for experience and managerial skills was included in the form of the variable, HED, measuring the years of schooling of the head of household. Our expectations concerning the negative sign on this variable is supported; the coefficient appears significant at the 6 to 12 percent level in the OLS equations (Table 3.4, Equation 1 and 2) and at the 16 percent level in the 2SLS equations (Table 3.5, Equation 1). As the household head acquires more schooling, the productivity of his time spent on off-farm activities as well as on farm managerial activities increases resulting in a reduction of time spent in the fields as a farm worker. If the husband's farm work time decreases, one would anticipate that the wife's farm time will increase to make up for the reduction in the husband's time. The positive coefficient on the husband's equation (HED)

TABLE 3.4
Ordinary Least-Squares Estimates of Farm Labor Supply Equations[a]

Variable	Head (1)	Head (2)	Wife(p)[b] (3)	Wife1 (4)	Wife2 (5)
Intercept	441.649	526.118	204.734	328.883	136.167
	(.0189)[c]	(.0041)	(.1146)	(.0837)	(.6523)
TFH	--	--	.5521	.453	.595
	--	--	(.0001)	(.0125)	(.0147)
TFW	0.057	--	--	--	--
	(.2634)	--	--	--	--
TFC	-0.049	-0.079	--	--	--
	(.2567)	(.0396)	--	--	--
TFO	0.056	0.082	0.0333	0.042	0.047
	(.0717)	(.0039)	(.0304)	(.0820)	(.1277)
CO4	-52.741	-62.282	11.244	16.184	-10.745
	(.0411)	(.0129)	(.5178)	(.5025)	(.8180)
AYC	-2.178	-10.422	13.548	14.089	-6.487
	(.8751)	(.4393)	(.0724)	(.1207)	(.8625)
MC714	38.446	61.753	-16.075	-21.341	2.050
	(.1844)	(.0263)	(.4007)	(.4867)	(.9587)
FC714	32.764	45.777	13.524	25.409	6.923
	(.3018)	(.1346)	(.5745)	(.4751)	(.8856)
HAGE	1.859	3.383	--	--	--
	(.5433)	(.2593)	--	--	--
HED	-27.854	-33.018	17.529	13.939	13.687
	(.1164)	(.0555)	(.1179)	(.4932)	(.5255)
OED	--	--	2.737	10.762	1.657
	--	--	(.8219)	(.5635)	(.9377)
AREA	-8.972	-7.256	-23.711	-38.029	-19.775
	(.5899)	(.6491)	(.0282)	(.0324)	(.4157)

(Continued)

TABLE 3.4 (Continued)

Variable	Head (1)	Head (2)	Wife(p)[b] (3)	Wife1 (4)	Wife2 (5)
AT	-.643	.133	-1.390	-.852	-.860
	(.4669)	(.8810)	(.0246)	(.4357)	(.4367)
PFI	.025	.290	-.005	-.003	-.008
	(.1648)	(.0951)	(.7111)	(.8763)	(.8292)
NW[d]	--	-80.872	66.286	54.817	71.153
	--	(.0271)	(.0319)	(.2372)	(.2742)
AW	--	--	-1.454	-2.228	-1.109
	--	--	(.4117)	(.4273)	(.8296)
$\bar{R}^2$	.203	.270	.322	.199	.357
	(.0340)[e]	(.0095)	(.0001)	(.0455)	(.0490)

[a]Dependent Variable: Average Farm Labor Hours Per day x 180 (days). Equations were estimated separately using OLS with 55 observations on head of households, 101 observations in the pooled sample of wives, and 55 observations on first and 34 observations on second eldest wife.

[b]Wife(p) indicates the labor supply function estimated from the pooled sample. There were 34 out of 56 households considered for which there were more than two wives present.

[c]Values in parentheses are probability values (p-values), i.e., Pr ($|t| > t^*$) where t* is the computed t-ratio. The p-value represents the smallest level of significance that would allow the null hypothesis to be rejected. The p-values here correspond to significance levels for one-tailed tests.

[d]NW also represents NW minus one in the female labor supply function.

[e]The values in parentheses represent p-values associated with an F-test of the null hypothesis based upon the unadjusted R^2. Adjusted R^2 values are reported however because they portray more accurately the goodness-of-fit.

TABLE 3.5
Two Stage Least-Squares Estimates of Family Members' Farm Labor Supply Equations[a]

Variable	Head (1)	Wife(p) (2)	Children (3)	Others (4)
Intercept	628.373[b]	-1305.801	-1249.051	518.967
	(.0023)	(.1805)	(.0050)	(.1364)
TFH	--	3.0795	1.1288	-.2808
	--	(.0651)	(.0489)	(.5923)
TFW	0.0812	--	.1005	-.0895
	(.3631)	--	(.3441)	(.3643)
TFO	-0.0253	-.0881	.0608	--
	(.6747)	(.0490)	(.5012)	--
TFC	-0.0219	-.0203	--	-.0375
	(.6453)	(.6424)	--	(.5142)
AYC	--	43.1831	-9.3557	--
	--	(.0847)	(.6910)	--
CO4	--	165.043	--	--
	--	(.1038)	--	--
MC714	39.2841	-198.696	-3.3508	-31.2018
	(.2275)	(.1286)	(.9525)	(.5725)
FC714	40.3972	-118.274	-51.7733	-13.4381
	(.2358)	(.1925)	(.4070)	(.8148)
NW	-119.434	--	--	--
	(.0234)	--	--	--
AT	.8457	-.4418	1.1183	1.2950
	(.4365)	(.6126)	(.3623)	(.3975)
PFI	.0307	-.0781	-.0097	.0298
	(.1122)	(.1446)	(.7667)	(.3853)
HAGE	1.1558	-5.0005	18.1660	--
	(.7476)	(.3986)	(.0001)	--

(Continued)

TABLE 3.5 (Continued)

Variable	Head (1)	Wife(p) (2)	Children (3)	Others (4)
HED	-27.4304	85.2674	32.1587	--
	(.1657)	(.0575)	(.3219)	--
OED	15.2303	--	-20.0301	49.5048
	(.4049)	--	(.4475)	(.0488)
AW	--	-.1284	--	--
	--	(.9578)	--	--
EXLTH	--	.3999	--	--
	--	(.2479)	--	--
NW-1	--	193.127	--	--
	--	(.0703)	--	--
AREA	10.0540	-16.0517	-22.0208	47.3581
	(.6902)	(.5461)	(.5534)	(.0079)
$\bar{R}^2$ [c]	.0304	.0539	.2375	.1895
	(.3572)	(.1782)	(.0006)	(.0031)

[a]Because each group involved a different sample size, unsophisticated 2-SLS estimates were obtained by (1) estimating the reduced form equation, (2) estimating an instrument for each dependent variable, and (3) applying least squares to the structural equation.

[b]Values in parentheses are p-values associated with asymptotic 't-tests' of the null hypothesis H_0: pi = 0.

[c]While goodness-of-fit measures such as R^2 and $\bar{R}^2$ are single-equation oriented and do not reflect well the fit of the system we report individual R^2's with asymptotic p-values as an indication of the model's adequacy.

in the wife's equations (3, 4 and 5 in Table 3.4 and 2 in Table 3.5) suggests that this type of substitution takes place. Consequently, with the head spending more time in managerial functions, the wives are likely to spend more time working in the fields, other things held constant.

The coefficient on purchased farm inputs, PFI, appears, as expected, positive in all the equations for husbands, and statistically significant at 9 to 16 percent level. The coefficient on animal traction hours, AT, also appears positive (in two out of three equations), but statistically not significant. The positive association between new agricultural inputs (PFI or AT) and the husband's labor supply to farming activities can be explained through the effects of such growth-promoting inputs on the productivity (value) of his labor time allocated to farm production. When farming starts becoming more profitable as a result of the introduction of new (modern) inputs in traditional agriculture, it is reasonable to expect that farmers will be motivated to expend additional time on farming activities.

Turning to the estimates of the wives labor supply equations (3-5 in Table 3.4), we find some interesting results that complement those obtained for the head's labor supply equations. Given the assumption of separate households, we estimated one equation using a pooled data set with 101 wives, and two other equations one each for the eldest (the senior most wife) and the second eldest wife. It is not surprising to see the highly significant coefficients on the head's farm time (TFH) and the 'other's' farm time (TFO). While many of the variables that were expected to capture the effect of children (and their age composition) on the labor supply decisions of wives did not possess significant coefficients, at least the signs on the coefficients satisfied our expectations.

Whenever there is a difference in sign between the coefficients in the first (eldest) wife (Wife 1) and the second eldest wife (Wife 2) equations (columns 4 and 5 of Table 3.4) the sign of the coefficient in the pooled equations agrees with the eldest wife ('wife1') equations. For example, the coefficient on the young children 7 to 14 years old (CO4) is 16.184 for 'wife1' and -10.745 for 'wife2'; and it is 11.244 for the pooled equation. These results may imply the dominant role played by the eldest wife in the household. We furthermore believe this suggests that substitution in child caring is taking place; the younger wives spend more time in child rearing. Of course, the eldest wife is well beyond the childbearing age and, hence, not likely to be raising any small children on her own.

Male and female children age 7 to 14 are observed to have opposite effects on the wife's labor supply (although not significant); males have a negative and females a positive effect on farm labor supply of wives. Older male children begin to assume more responsibility for work on the farm and as the number of older male children increases, the likelihood of teen-aged males being present in the home increases, and their time is substituted for the wife's time, especially for the eldest wife's time. In contrast, older female children acquire more responsibility for household activities. Hence, as the number of female children increases, the likelihood of the presence of teen-aged females increases, and their time is substituted for that of the wife in the household, freeing the wife for farming.

The negative sign on and significance of the coefficient of farm size, AREA, is disconcerting. It is difficult to offer any reasonable explanation of this result. We note that the 2SLS estimates of the coefficient on farm size (AREA) are insignificant in all but the fourth equation for 'others' where it possesses a positive sign (Table 3.5). It seems doubtful if simultaneous equation bias can be considered as the source of these unexpected results. Animal traction is significant in the pooled sample equation, implying that an increase of one hour of animal traction use on the farm results in a decrease 1.39 hours of wives' farm time, on average. This result is consistent with that obtained in the household equations where the effect of this variable is significant and positive in the wife's equations.

The number of wives, NW, is strongly significant and with a positive sign. The result indicates that there is substitution and work sharing among wives; the addition of one wife in the household releases an individual wife, on average, to do about 66 hours more farm work during the season, other things being equal.

Summing up, the results of the household head's farm labor supply equations indicate that there is little substitution among the household's adult members with respect to the labor time for work in the field, this is due to the exogenously (weather) determined farm activity that must be performed in a given period of time. Some substitution is implied, however, between the number of wives and the husband's time devoted to farm work. The children's economic contribution to the household was revealed by a significant and negative coefficient of the adult children's farm time, and a positive coefficient on the number of younger children. The household head's schooling shows, as expected, a negative effect on the head's farm labor time, while the use of animal traction and purchased farm inputs show positive effects.

The results of the wives' farm labor supply equations generally complement those of the husbands. The estimated coefficients on children appear mixed; although most of the coefficients appear insignificant, their signs are generally consistent with our expectations. The coefficient on the number of children (0-4 years old) is negative in the second wife's equation, while positive in the first (senior most) wife's equations, suggesting for some substitution between the wives in child caring. The older male children show a negative effect, while the older female children, a positive effect, on the wife's farm labor time, implying substitution possibilities for children across the household and farm activities. The result on the number of wives indicate that there is substitution and work sharing among wives--a phenomenon of significant economic importance in polygynous African societies. Another interesting result is that the effect of the husband's education is to significantly increase the wife's farm labor time. The use of purchased farm inputs and animal traction do not seem to show any statistically significant effect on the wife's farm labor time, however. The effect of the farm size appears, surprisingly, negative and significant--a result difficult to explain.

III. HOME PRODUCTION LABOR SUPPLY, MARGINAL PRODUCTIVITY AND THE DOLLAR VALUE OF WORK-AT-HOME OF WIVES

Estimates of the labor Supply and Marginal Productivity Equation

Table 3.6 provides descriptive statistics for the variables as well as the definitions of the variables used in the analyses.

Estimates of the labor-supply function for both the pooled sample of 102 wives and the subsample of the 55 eldest wives are given in Table 3.7. The results of estimating the marginal-productivity equation[3] are presented in Table 3.8. The results of bootstrapping the coefficients of the marginal-productivity function (Table 3.8) are consistent with the results obtained from the ordinary least-squares (OLS) estimation of the wives' labor-supply function for work-at-home (Table 3.7).

The results presented in the two tables suggest that several variables influence the marginal productivity of the wives' work-at-home. For the function estimated from the pooled sample of 102 wives, the age of the wife (AGE), the hours of animal traction employed on the farm (AT), and the number of other wives in the household (NW-1) all appear highly significant statistically in determining the wife's productivity in home production. As expected, marginal productivity declines with the age of the wife and with increases in the number of wives in the household. While it is likely that marginal productivity

TABLE 3.6
Sample Means and Standard Deviations of Variables

Variable[a]	All Wives (N = 102)		Eldest Wife Subsample (N = 55)	
	Mean	SD[b]	Mean	SD[b]
AGE	38.6961	411.902	44.7037	340.049
AYC	2.5588	34.935	2.9074	29.403
C04	2.1662	27.863	1.8704	18.588
MC714	1.7745	25.017	1.4815	15.225
FC714	1.3922	19.699	1.2222	12.590
AT	36.6078	724.538	23.5556	392.274
AREA	6.0083	76.951	5.0099	45.615
HED	.8911	31.847	.5000	17.457
NW-1	1.7843	22.196	1.4074	13.417
FCH3	66.4286	1762.170	66.8650	1300.610
LNW2	1.3603	13.854	1.3113	9.725

[a]Exogenous variables used in analysis:

AGE = age of wife
AYC = age of youngest child in household
C04 = number of children between the ages of zero and four
MC714 = number of male children between the ages of seven and fourteen
FC714 = number of female children between the ages of seven and fourteen
AT = hours of animal traction employed in the farm
AREA = size of farm in hectares
HED = number of years of schooling of head of household
NW-1 = number of head's wives in household minus one
FCH3 = number of hours of household labor supplied by daughters of head's wives to household category three
LNW2 = natural log of the implicit wage rate of wife (constant across all wives in a given household). The implicit wage rate for wives for each household was based upon the estimate of the elasticity of female labor input on farm production (the production function used to estimate the implicit wage rate had the Cobb-Douglas type production function).

[b]Standard Deviation.

TABLE 3.7
The Estimated Coefficients of the Wives' Home Labor Supply Function (Ordinary Least-Squares)

Variable[a]	All Wives (N = 102)			Eldest Wife Subsample (N = 55)		
	Coefficient	SD[b]	t-ratio	Coefficient	SD[b]	t-ratio
Constant	1693.0100	176.1411	9.6117	1633.6400	221.7724	7.3663
AGE	-5.3750	1.8202	-2.9350	-6.4313	2.6848	-2.3954
AYC	5.7785	12.3819	.4667	10.0779	3.1421	3.2074
C04	25.4774	17.7151	1.4382	54.4221	8.8029	6.1823
MC714	5.0927	19.8312	.2568	21.3569	28.4557	.7505
FC714	-17.7466	12.7432	-1.3926	-23.8215	32.3481	-.7364
AT	1.8701	.6773	2.7613	1.1802	1.0351	1.1402
AREA	10.8738	9.2873	1.1746	22.7340	14.1385	1.6079
HED	.6058	10.5389	.0575	-8.6103	17.9155	-.4806
NW-1	-82.2965	29.2474	-2.3138	-94.3805	40.0234	-2.3581
FCH3	-.2357	.1686	-1.3980	-.3373	.2046	-1.6491
LNW2	-300.0000	--	--	-300.0000	--	--
$R^2 = .3037$	$\bar{R}^2 = .1985$	$F(11, 91) = 3.6082$;		$R^2 = .3567$	$\bar{R}^2 = .2132$	$F(11, 44) = 2.2179$

[a]Exogenous variables as defined in footnote a, Table 3.6.
[b]Standard Deviation.

increases with age, such increases occur during a short interval of time, too short to be detected given the sample sizes studied.

It is suspected that the negative effect of age is due in part to the debilitating effects of the harsh environment (including malnutrition and the conspicuous absence of health-care facilities) on the physical capacity of rural women in Burkina Faso. And this is likely to be true throughout much of rural West Africa. Any increase in productivity due to the effect of experience that would be correlated with increases in age is likely to be short-lived and outweighed by decreases in productivity due to physical deterioration.

The individual wife's marginal productivity in home production declines with increases in the number of other wives quite clearly because some of the housework is distributed across a larger number of female workers. In the polygynous family structure, while each wife is placed in charge of her own household or subhousehold, some home production is performed jointly by several wives, for example, the task of milling grain. It is reasonable to assume that the presence of multiple wives will have a significant effect on the household's labor supply decisions and on the estimated marginal productivity of an "average" wife. Furthermore, as the household ages, the head of the household tends to acquire additional wives, and, with an increase in the number of wives, the responsibilities and the tasks performed by wives also undergo some changes. Consequently, the increase in the number of wives that the head acquires over the life cycle of the eldest (the most senior) wife may have an important influence on the value of the latter's contribution to the household.

The two farm variables that appeared significant determinants of the wife's value of home production are the use of animal traction on small farms and the wife's implicit wage rate. Animal traction, the donkey-drawn or ox-drawn plows, planters and weeders, is a labor-saving device. It needs to be noted that in recent years there has been a growing interest among the international agencies, as well as the national governments in several African countries, in promoting the animal traction technology in the region as a device to break the labor constraint and to raise farm production.[4] The use of animal traction reduces the demand for the wives' labor time on the farm. (Animal traction is generally operated by the older male children.) Hence, the women are able to spend more time in home-production activities at times when they can be more productive. The positive and statistically significant coefficient

TABLE 3.8
Estimates of Coefficients of the Marginal Productivity Function For Wives' Work-at-Home

	All Wives (N = 102)			Eldest Wife Subsample (N = 55)		
Variable[a]	Estimated Coefficient[b]	Implied t-ratio[c]	Robust t-ratio (MED/MAD)[d]	Estimated Coefficient[b]	Robust Implied t-ratio[c]	t-ratio (MED/MAD)[d]
Intercept	5.6434	13.7610***[e]	23.7395***	5.4455	9.1985***	13.7075***
AGE	-.0179	-2.5571***	-3.6170***	-.0214	-1.1889	-2.7368***
AYC	.0193	.4520	.3655**	.0036	.5153	1.1685
CO4	.0849	1.2322	1.6726**	.1814	1.8176**	2.7145***
MC714	.0170	.2439	.2349	.0712	.6526	.9924
FC714	-.0592	-.7345	-1.1440	-.0794	-.6307	-.8828
AT	.0062	2.6955***	4.0667***	.0039	1.0006	2.0000**
AREA	.0362	1.2440	1.5708***	.0758	1.3908	1.7698
HED	.0020	.0545	.0325	-.0287	-.4165	-.6265
NW-1	-.2743	-2.7596***	-3.4607***	-.3146	-2.0167**	-3.1610***
FCH3	-.0008	-1.1429	-2.2500***	-.0011	-1.2222	-2.6000***
$\ln T_{Z3}$	-.0033	-6.6000***	11.0000***	-.0033	-4.7143***	-8.2500***

[a]Dependent variable = log of marginal-value product of work-at-home, $\ln h_2^*$.

Independent variables are:

AGE = age of wife

AYC = age of youngest child in household

C04 = number of children between the ages of zero and 4

MC714 = number of male children between the ages of 7 and 14

FC714 = number of female children between the ages of 7 and 14

AT = hours of animal traction employed on the farm

AREA = size of farm in hectares

HED = number of years of schooling of head of household

NW-1 = number of head's wives in household minus one

FCH3 = number of hours of household labor supplied by daughters of head's wives to household category three

$\ln T_{Z3}$ = natural log of time allocated to home production activities by wife during the entire year, measured as 360 days

[b]Estimated coefficients are obtained from the relations expressed in Equation 9 and the bootstrapped sampling distributions (number of bootstrap replicate samples was 100).

[c]The implied t-ratio is the pivotal quantity constructed as the ratio of the estimated coefficient to its bootstrap standard error.

[d]Robust t-ratios are defined as the ratio of the coefficient estimate based on the median (MED) of the bootstrap estimates to the estimate of the coefficient's standard error, which is based on a robust estimate of standard error called the median absolute deviation (MAD). These estimates are provided as a robust alternative to the conventional t-ratios based on the normality assumption.

[e]Statistical significance: *** = 1% level, ** = 5% level, and * = 10% level.

on the animal traction hours variable, AT, confirms this expectation (Tables 3.4 and 3.5, all-wives equations). The introduction of animal traction technology, therefore, may not imply overall less work for the wives.

The other farm variable, the wife's implicit wage rate, f_3^* is based on the output elasticity of the wife's labor time input in farm production estimated using the Cobb-Douglas type production function. The coefficient on the variable f_3^* appears, as expected, negative and highly significant statistically. In the dominantly labor-intensive production system that is found in the traditional agricultural setting of the region, the amount as well as the value of the wife's time allocated to farming activities is often significant. As the value of the wife's labor time, or the marginal productivity of the wife in agricultural production, rises, the wife's time in home production tends to become relatively more expensive; and, hence, it is reasonable to expect that the wife will allocate less time to home production, ceteris paribus.

The effects of other variables included in the estimating model, that is, the number of children between the ages of zero and 4 years present in the household, CO4, the size of the farm (in hectares), AREA, and the number of hours of work spent in home production by the female children, FCH3, appeared marginally significant statistically (Table 3.8). Children under 4 years of age require more time and attention from the mother, hence they exert a positive effect on the mother's marginal productivity in home production. The value of the mother's time in home production, therefore, tends to rise as the number of children of this age in the household increases. However, it is also true that when adult unmarried female children are present in the household they share in work-at-home. Although, since most daughters in these households marry at an early age, the effect observed through the variables FCH3, that is, the young female home-production time, might have resulted in rendering the effect of this variable statistically weak as indicated by the OLS estimates of the productivity equation. The result for this variable obtained through the bootstrapped productivity equation, as discussed below, needs to be noted, however.

Overall, results similar to those found with the pooled estimation are obtained for the bootstrapped marginal-productivity equation based on the sample of observations for the 55 eldest (senior) wives (Table 3.8). Statistical significance levels are generally higher for coefficients of the equation based on the eldest wife subsample. However, note the difference in the t-ratios associated with the coefficient on the log of the wife's implicit wage rate, f_3^*. The coefficient is statistically less significant in the equation for the eldest wife, a result that is consistent with the values of the F-test of the parameter constraint imposed in the work-at-home function (Table 3.7).

The interesting thing to note, however, is that the use of the conservative robust t-tests resulted in some changes in the levels of significance of the estimated coefficients on variables of the wife's marginal productivity equations (Table 3.8). The most noteworthy change appeared for the coefficient on the number of hours of household labor supplied by the adult female children (the daughters of the household head) to home production. In the marginal-productivity equation based on the pooled sample of 102 wives, the coefficient on the variable FCH3 (the female children's labor time in home production) rose in significance from the 0.15 level to the 0.025 level. A similar change appeared in the eldest wives equation. These results suggest that the adult female children's labor time variable is a much more significant determinant of the wife's marginal productivity in home production than what may be suggested by the results of the standard t-tests. Similarly, the significance of the coefficients on the age of the youngest child, AYC, the number of younger children 0-4 years old, CO4, and the household's farm size, AREA, increased from the 10% level given by the standard t-tests to the 5% level or smaller indicated by the robust t-tests for both the pooled equation and the eldest wives equation (Table 3.8). The differences between the standard and the robust t-tests, therefore, suggest there is reason to doubt the implications of the standard t-tests.

Overall, the robust t-tests would imply a larger number of significant determinants of the wives' marginal productivity in home production and, hence, the wives' value of work-at-home than would be inferred from the results of the standard t-tests. Thus, we may also conjecture that these results place on firmer ground the interpretations and the inferences drawn about the estimates of value obtained in this study and serve to strengthen our proposition about the determinants of the wives' marginal productivity in home production.

Estimates of the Dollar Value of Wives' Work-at-Home

The estimates for the dollar value of work-at-home performed by an individual wife in an average household are presented in Tables 3.9 through 3.12. The estimates of value for the wife's home-production services are based on the pooled sample of 102 wives, as well as the (senior) eldest wives subsample of 55,

TABLE 3.9
Estimates of the Value[a] of Home Production With Given Combinations of Farm Size, Number of Wives and Younger Children[b]

Key Variables		Value	Mean	Median	
AREA	NW-1	V	$\bar{V}$[c]	M(V)[d]	SD(V)[e]
4	1	471	502	447	172
4	2	358	377	335	125
4	3	272	298	267	128
6	1	506	559	498	302
6	2	385	419	385	134
6	3	292	321	303	102

Value of other variables: C04 = 2, WH3 = 400, FC714 = 3

Key Variables		Value	Mean	Median	
AREA	NW-1	V	$\bar{V}$[c]	M(V)[d]	SD(V)[e]
1	1	462	527	455	248
1	2	471	502	447	172
1	3	480	523	476	181
2	1	351	381	359	161
2	2	358	377	335	125
2	3	365	408	377	132

Value of other variables: AREA = 4, WH3 = 400, FC714 = 3

[a]All values are in U.S. Dollars (1 U.S. $ = 225 CFA francs). All Wives (N = 102)

[b]Bootstrapped confidence intervals, using data from all wives (N = 102). We have reported several descriptive measures of the bootstrap distributions for the value estimates V, to give some idea of the shape of these distributions.

[c]$\bar{V}$ = the average of 100 simulated values of V.

[d]M(V) = the median of 100 simulated values of V.

[e]SD(V) = the standard deviation of the 100 simulated values of V.

TABLE 3.10
Estimates of the Value[a] of Wives' Home Production With Given Combinations of Key Variables

Key Variables		Value	Mean	Median	
NW-1	FC714	V	$\bar{V}^c$	M(V)[d]	SD(V)[e]
1	1	530	503	542	220
1	2	500	533	472	252
1	3	471	502	447	172
2	1	477	486	453	221
2	2	380	445	391	248
2	3	292	321	303	102

Value of other key variables: AREA = 4, C04 = 2, WH3 = 400

Key Variables		Value	Mean	Median	
NW-1	AGE	V	$\bar{V}^c$	M(V)[d]	SD(V)[e]
1	20	674	771	680	326
1	30	563	610	528	235
1	40	471	543	511	252
1	50	394	417	390	173
1	60	329	366	324	180

Other variables: AREA = 4, C04 = 2, WH3 = 400, FC714 = 2

[a]See footnotes a-e of Table 3.9.

given a variety of combinations of the key determinants of marginal productivity. In particular, the effects of the size of the farm (AREA), the number of the other wives of the household head (NW-1), the number of young children between the ages of zero and 4 (CO4), the number of the female children between the ages of 7 and 14 (FC714), and the age of the wife (AGE) on the estimated value of the wife's home production have been explored (Tables 3.9, 3.10, and 3.11). Additionally, the interaction between the age of the wife (AGE) and the number of the other wives of the household head (NW-1) is examined through a simulation to gain some insight into their dynamic effects on the value of the wife's economic contribution to the household over her life cycle (Table 3.12).

TABLE 3.11
Estimates of the Value[a] of Home Production For the Eldest Wives Subsample With Given Combinations of Farm Size, Number of Wives and Number of Children Age 7-14 Years

Key Variables		Value	Mean	Median	
AREA	NW-1	V	$\bar{V}^c$	$M(V)^d$	$SD(V)^e$
4	1	390	409	358	218
4	2	284	345	285	207
4	3	207	231	212	117
6	1	452	543	468	299
6	2	330	425	351	235
6	3	241	263	221	135

Value of other key variables: FC714 = 2, C04 = 2, WH3 = 400

Key Variables		Value	Mean	Median	
NW-1	FC714	V	$\bar{V}^c$	$M(V)^d$	$SD(V)^e$
1	1	456	556	484	311
1	2	421	599	421	320
1	3	207	231	212	117

Values of other variables: C04 = 2, WH3 = 400

[a]See footnotes a-e of Table 3.9, except that for this subsample N = 55, the number of eldest wives.

The estimates of value of home production clearly reveal that the extent of the economic contribution that farm wives make to the rural household depends on the characteristics of the household. As evident from Tables 3.9 and 3.10, the value of home production increases, as expected, with the number of young children of 0-4 years of age (CO4) and the size of the farm (AREA), while it decreases with increases in the number of wives (NW-1) and in the number of the female children 7-14 years old (FC714). All of the changes in value occur for quite obvious reasons. For example, the estimated mean value of work-at-home for a wife, given one other wife, two young children (0-4 years old), and a farm size of four hectares (9.2 acres) is \$471 (at the 1980 exchange rate of 225 CFA francs

TABLE 3.12
Estimates of the Value[a] of Home Production For the Eldest Wife Simulating the Dynamic Interaction Between Age and the Number of Other Wives

Key Variables		Value	Mean	Median	
AGE	NW-1	V	$\bar{V}^c$	$M(V)^d$	$SD(V)^e$
20	0	913	1,154	887	963
30	1	538	730	577	612
40	2	317	457	324	636
50	3	187	207	188	91

Value of other key variables: AREA = 4, C04 = 2, FC714 = 2

[a]See footnotes a-e of Table 3.9, except that for this subsample N = 55, the number of eldest wives.

= U.S. $1). This value of home production in an absolute term may not appear large. But considering that the estimated average annual farm income of households generated through crop production was $770, a value of $471 generated through work-at-home assumes significance (annual farm income equals gross revenue from all crops grown on the farm minus purchased farm inputs). The significance of the value of home production, in this context, lies in the fact that this value (of home production) equals 61% of the level of family income derived from the farm.[5] And, for the poor rural households, the wife's economic contribution through home production to family income and welfare is quite substantial. The value of home production, however, drops, as expected, to $358 (which equals 46% of the level of the family farm income) with two additional wives present; and, finally, to $292 (which equals 38% of the family farm income) with three additional wives present (Table 3.9).

An increase in the number of female children 7 to 14 years old also results in decreasing the value of the wife's contribution mainly because the female children's work-at-home substitutes for the work done by wives. As revealed by the estimates (Table 3.10), an average wife's contribution of $530 with one additional wife and one female child 7-14 years old declines to $471 with three female children 7-14 years old.

The effects of increases in the wife's age (AGE) that seem to represent, among other things, the effects of the physically demanding, harsh rural way of life are reflected in the significant declines in the value of work-at-home for an average wife in a household with only one other wife (Table 3.10). From age 20 to age 60, the average value of home production contributed by the wife is nearly cut in half; at age 20 the value is estimated at $674, and at age 60 the value declines to $329. The estimating model predicts that the wife at age 60 is still individually supplying household services, the dollar value of which would still equal about 43% of the level of family farm income, on average. It is true that this estimate does not adequately reflect the entire dynamics of the household during this 40-year span and probably overstates the woman's economic contribution. However, what is abundantly clear from the results is that this contribution is not insignificant.

Table 3.11 dramatically illustrates the significant differences between the contribution of the eldest wife and that of the "average" wife as presented in Tables 3.9 and 3.10. The value of home-produced services of the eldest (the most senior) wife is, as expected, consistently smaller than the average wife. The important point to note is that although the eldest wife's economic contribution to home production is relatively smaller than the average (younger or junior) wife, the dollar value of her work-at-home ranges from $207 to $456, and this still amounts to 27%-59% of the level of the family farm income.

An attempt to capture important dynamic aspects of the household and their influence on the value of home production is summarized (Table 3.12) for a typical pattern of corresponding changes in the number of the other wives of the household head (NW-1) and the age of the head's eldest wife (AGE). In contrast to the results presented in Tables 3.9 and 3.10 for an "average" wife, the substitution effect due to increases in the number of other wives of the household head, coupled with the decline in physical capacity that may accompany the aging process, reduces the value of home production for the eldest wife from a high of $913, at age 20, to a low of $187, at age 50, and with three other wives present (Table 3.12). We may sum up, then, that while other determinants of marginal productivity are surely changing over the life cycle as well as those that are chosen in this study, the results of this study are strongly suggestive of the important economic contributions that wives make to family income and welfare.

Summary and Conclusions of Wives' Home Production

The results of the study of wives' labor supply to home production and the value of her work-at-home can now be summed as follows.

The variables of the estimating model that emerged as significant determinants of the wife's marginal productivity in home production are the wife's age, with a negative effect; the number of hours of animal traction used in farm production, with a positive effect; the number of other wives of the household head, with a negative effect; and the wife's implicit farm wage, that is, the marginal productivity of the wife's time in farm production, with a negative effect.

The use of the robust t-tests implies greater significance for the estimates of the coefficient in the marginal-productivity equation: the statistical significance of the coefficients on the female children's labor time use in home production, the age of the younger children in the household, "increased" from 10%-15% to 5% or lower over the standard t-tests. These results implied a larger number of significant determinants of the wives' productivity in home production than were implied by the standard tests.

The estimated value of the wife's work-at-home has been shown to vary significantly with the economic characteristics of the household; from $292 to $471 (which equals 38%-61% of the level of the family farm incomes), according to one set of point estimates, and from $187 to $913 (which equals 25%-118% of the level of the family farm income), according to another set of estimates. The value of the wife's home production tends to decrease with the number of other wives of the household head and the number of the female children, while it tends to increase with the number of younger children, 0-4 years, and with the size of the farm. The effect of the wife's age is also to decrease the value of her work-at-home. The results also appear to be strongly suggestive of the importance of certain dynamic interactive effects of the household characteristics on the value of home production over the wife's life cycle.

Overall, the estimates of the value of home production clearly demonstrate the significant economic contribution that the African rural women make to family income and welfare--indeed to national income and welfare and, hence, to economic growth.

IV. LABOR SUPPLY, MARGINAL PRODUCTIVITY AND THE VALUE OF CHILDREN'S HOME PRODUCTION

Labor Supply and Marginal Productivity Equations

Table 3.13 provides descriptive statistics for the variables as well as the definitions of the variables used in the analyses.

The estimated coefficients of the adult children's work-at-home function are presented in Table 3.14, while estimates of the marginal productivity equation are presented in Table 3.15.

The results of estimating the work-at-home function (Table 3.14) and the marginal productivity equation (Table 3.15), are

TABLE 3.13
Definitions, Means and Standard Deviations of Variables Used in the Statistical Analysis

Variable and Definition		Mean	Standard Deviation
CO4	= the number of children between the ages of 0 and 4	2.0000	2.1478
MC714	= the number of male children between the ages of 7 and 14	1.2857	1.0988
FC714	= the number of female children between the ages of 7 and 14	1.0634	0.9311
AGE	= the age of the child	21.8700	7.6548
AGE2	= the square of the child's age	536.0900	402.6955
NW	= the number of wives of the head of the household	2.4921	1.3183
CHT	= total time spent in work at home by child	481.5612	391.5795
LVCTF[a]	= log of daily implicit wage rate	3.4255	0.6667

[a]LVCTF ($\ln h_2$ of equation 8, see Model in the Appendix of this chapter) is based upon the output elasticity of child labor input in the farm production function, estimated as a Cobb-Douglas form with constant returns to scale.

TABLE 3.14
OLS Estimates of Coefficients of the Work-at-Home Function for Children Ages 15 and Above

Variable	Coefficient	Standard Deviation	t-ratio	PROB > t
Inter.	681.380	527.849	1.291	0.2022
CO4	107.474	19.628	5.476	0.0001
MC714	-11.509	43.197	-0.266	0.7909
FC714	12.253	54.673	0.224	0.8235
AGE	-3.444	39.755	-0.087	0.9313
AGE2	0.120	0.752	0.159	0.8740
NW	-17.239	34.555	-0.499	0.6199
LVCTF[a]	-104.776	64.368	-1.628	0.1093

R^2 = 0.3943 Adj R^2 = 0.3172 F(7,55) = 5.115
PROB > F = 0.0002 RMSE = 323.567 Condition Index = 13.456

[a]Coefficient on LVCTF restricted to -145, F(1,55) = 0.3905, PROB > F = 0.5346.

quite consistent with respect to the explanatory variables that are significant determinants of either the work-at-home, or the marginal productivity in home production. As evident from the statistical results, several variables operating at the household and farm levels seem to influence the children's marginal productivity in home production and, hence, the value of their work-at-home. The presence of young children between the ages of 0 and 4 years (CO4) has the most statistically significant effect on the time spent on work-at-home and on the marginal productivity and, ultimately, on the value of a child's household production. According to the estimates in Table 3.14, one additional child in this 0 to 4 age group induces an adult child to supply an additional 107 hours for home production activities over the course of one year, other variables constant. In both the work-at-home function and the marginal productivity equation the coefficient of the variable, the number of younger children (0 to 4 years old, CO4), carries the expected sign, and it is statistically significant at a level of five percent or less. This result may imply that: (a) the presence of young children in the household raises the value of

TABLE 3.15
Estimates of the Coefficients of the Marginal Productivity[a] of Work-at-Home Function for Adult Children 15 Years and Over

Variable	Coefficient	Standard Deviation	t-ratio	BST Mean	BST Med
CO4	0.7412	0.2724	2.7210	0.8122	0.7616
MC714	-0.0794	0.1685	-0.4712	-0.1071	-0.0915
FC714	0.0845	0.2187	0.3864	0.0804	0.0852
AGE	-0.0238	0.1051	-0.2265	-0.4860	-0.0262
AGE2	0.0008	0.0023	0.3478	0.0013	0.0008
NW	-0.1189	0.2168	-0.5484	-0.1435	-0.1237
CHT	-0.0069	0.0027	-2.5556	-0.0076	-0.0071
Inter.	4.6992	1.6835	2.7913	5.2223	4.8024

[a]The dependent variable is log of marginal productivity, $\ln h_2$. Estimates are based upon Equation 9 and the bootstrapped sampling distributions. The number of bootstrap relicate samples was 100.

time adult children spend in home production, (b) an important role of the young adult children, particularly of the female children, in home production is child care, and (c) the children's time spent on child care reflects a substitution of their time for that of their parents whose marginal productivity is likely to be higher for other activities.

The coefficient of the total home production time of adult children (CHT) appears negative as expected; the home production function has been assumed to exhibit decreasing marginal productivity and this empirical result would seem to support that assumption. We note that the coefficient on home production time is obtained, as are all of the other coefficients in the marginal productivity equation, under an assumption of equality of the marginal value products in home and farm production.

Other determinants, such as the number of male and female children between the ages of 7 and 14 (MC714 and FC714), the age of the adult child (AGE), and the number of wives of the household head (NW), do not appear to have any statistically significant impact either on the children's time supplied to home production, or on the marginal productivity of their work-at-home.

The farm variable, the children's implicit daily wage rate (LVCTF), included in the estimating model is based upon the output elasticity of the adult children's labor time input in farm production estimated using the Cobb-Douglas type production function. The coefficient on the variable (LVCTF) is, as expected, negative in the work-at-home function and statistically significant at the eleven percent level of significance. Within the highly labor intensive production system that characterizes the traditional agricultural setting of the region, the amount, as well as the value of the children's labor time allocated to farming, is, like that of their mothers often significant. As the value of the children's time on the farm, the marginal productivity of children in agricultural production rises, the adult children's time spent in some of the household activities tends to become relatively more expensive. Consequently, it is reasonable to believe that the children will tend to allocate relatively less time to home production activities, other things constant.

Estimates of the Dollar Value of Children's Work-at-Home

The results of estimating the dollar value of adult children's work in home production are presented in Table 3.16. The estimates of the value of home production are based upon the sample of both the male and the female children in the household, given combinations of the significant variables determining the productivity of children in home production. For example, considering that the household consists of three groups of family members, the household head, the wives and the adult children, plus two children between the ages of 0 and 4 years, one male child and one female child between the ages of 7 and 14 years, an adult child contributes, on the average, approximately $190 to the family's annual income (Table 3.16). The mean value of an adult child's work-at-home increases with increases in the age of the child, all else held constant, although the increase is not significant. It is worth stating here that since many of the children eventually marry and start households of their own, the participation of children in their parent's household activities will, other things constant, tend to decline eventually as they grow older.

The other significant determinant of the value of children's work-at-home is the number of very young children between the ages of 0 to 4 years. As evident from the estimates in Table 3.16, the annual value of work performed by 21 year old

TABLE 3.16
Estimates of the Value[a] of Work-at-Home of Adult Children Age 15 and Above with Given Combinations of Age and the Number of 0-4 Years Old Children in the Household[b]

Value	Mean	Median	Standard Deviation	LB[c]	UB[c]	Age	CO4
190	324	176	207	98	868	15	2
190	243	171	247	108	726	17	2
193	232	185	185	119	616	19	2
196	280	178	338	117	628	21	2
325	351	209	502	103	1197	21	4

[a]All values in US $ = 225 CFA francs.
[b]Other characteristics of the household: NW = 2.5, CHT = 484, MC714 = 1, FC714 = 1.
[c]LB and UB represent the 90 percent confidence lower bounds (LB) and upper bounds (UB) based upon the distribution of pseudo estimates of the value obtained through bootstrapping.

children, for example, increases from $196 to $325, an increase of 71 percent, as the number of younger children (0-4 years old) increases from 2 to 4, other variables remaining constant. This result was expected since the presence of very young children in the household implies greater need for child care work and, consequently, more work-at-home for the relatively old children, particularly the female children. The dollar value of children's home production equals 25 percent to 42 percent of family income derived from the farm. For the poor rural households, the contribution of rural children through home production is obviously significant. If we add the estimated value of work performed by the children in farm production (which is $86 annually) to the value of their home production ($190-$325), the overall economic contribution of children to family income and welfare will assume even greater importance.

Summary and Conclusions of Children's Home Production

The variables in the econometric model that emerged as significant determinants of the children's productivity in home production are the number of young children in the household

(CO4), with a positive effect, and the children's implicit farm wage rate (LVCTF), with a negative effect. As in the case of wives, the result of a negative coefficient on the total home production time of children supported the hypothesis of decreasing marginal productivity in home production. The effects of other variables such as the age of the (adult) child and the number of wives per household head statistically appeared weak.

The results of estimating the dollar value of children's home production varies with the economic characteristics of the household; from \$190 or \$196 with 2 children between the ages of 0 and 4 years, to \$325 with 4 children in the age of 0 to 4 years. This means that the value of children's work-at-home stands roughly between 25 percent and 42 percent of the level of the family farm income. The mean value of work-at-home increases with increases in the age of the children, other things constant.

In summary, quite clearly the children are an important source of income for the household and are clearly valuable investments to the parents. The decision to raise large numbers of children and to maintain extended families appears quite rational in light of the value or potential value of the income stream that the children can generate over their life cycle. Parents may not be directly calculating the "dollar" value of their children's contributions when deciding to produce and invest in children in the same way they may probably be doing while deciding to produce other goods. Nevertheless, they implicitly make rational calculations of the relative cost and the relative benefit, and these calculations have a significant bearing upon the parents' decisions about the family size and about the investment in the quality of their children, no matter how poor the parents may be.

APPENDIX TO CHAPTER 3: THE MODEL

Following Gronau we develop a model of home production that will lead to a computation of the value of women's work in home production.[6] We assume that the household seeks to maximize welfare (utility),

$$U = U(Z_1, Z_2, M), \qquad (1)$$

as a function of three types of goods consumed. Z_1 represents goods produced in the home by males, in particular the head. Z_2 represents goods produced in the home by females, namely the wife. All other goods consumed, M, are purchased in a marked at a price P_M. The goods Z_1 obey the production relation

$$Z_1 = g(X, T_{Z1}), \tag{2}$$

where X = other goods used by the head in home production, and T_{Z1} = time allocated by head to home production. Goods Z_2 = are defined by the production relation[7]

$$Z_2 = h(\alpha Q, T_{Z2}), \tag{3}$$

where Q = output from the farm, α = proportion of total farm output used by wife in home production $(0 < \alpha < 1)$, T_{Z2} = time allocated by wife to home production. Farm output, Q, is obtained according to the production relation

$$Q = f(F, T_{Q1}, T_{Q2}), \tag{4}$$

where F = purchased farm inputs, T_{Q1} = time allocated by household head to farming, and T_{Q2} = time allocated by wife to farming.[8] In addition to the production constraints (2) through (4), utility maximizing choices of Z_1, Z_2 and M are made subject to three additional constraints:

$$T_1 = T_{Z1} + T_{Q1}, \tag{5a}$$

$$T_2 = T_{Z2} + T_{Q2}, \tag{5b}$$

$$T_M M + P_F F = P_Q(1 - \alpha)Q, \tag{5c}$$

where T_i = total time (net of maintenance), i = 1, 2, P_M = price of purchased consumption goods, P_F = price of purchased farm inputs, P_Q = price of farm output, and $(1 - \alpha)$ = proportion of farm output marketed.

Equating $T_{Q1} = T_i - T_{Zi}$ for i = 1, 2 from (5a) and (5b) and substituting into (4) the first-order conditions are given by:

$$\delta/\delta Z_1: \quad U_1 + \lambda_1 = 0 \tag{6a}$$

$$\delta/\delta Z_2: \quad U_2 + \lambda_2 = 0 \tag{6b}$$

$$\delta/\delta M: \quad U_3 - \lambda_4 P_M = 0 \tag{6c}$$

$$\delta/\delta Q: \quad -\lambda_2 h_1 + \lambda_3 + \lambda_4 P_Q(1 - \alpha) = 0 \tag{6d}$$

$$\delta/\delta F: \quad -\lambda_3 f_1 - \lambda_4 P_F = 0 \tag{6e}$$

$$\delta/\delta \alpha: \quad -\lambda_2 h_1 Q - \lambda_4 P_Q Q = 0 \tag{6f}$$

$$\delta/\delta T_{Z1}: \quad -\lambda_1 g_2 + \lambda_3 f_2 = 0 \tag{6g}$$

$$\delta/\delta T_{Z2}: \quad -\lambda_2 + \lambda_3 f_3 = 0 \tag{6h}$$

Solution of the first-order conditions (6a) through (6h) will lead to the major implication that, in equilibrium, the wife will allocate time between work in the home and farm production so that the value of her marginal product in home production is equal to the value of her marginal effort on the farm. That is, from equations (6g), (6c), (6d), and (6e),

$$MVP_{farm} = (h_1\alpha)f_3 = [1 + (P_Q/P_F)f_1(1 - \alpha)]h_2 = MVP_{home}, \tag{7}$$

where $(h_1$ is the implicit wage rate for the wife's farm labor, f_3 is the marginal productivity of the wife in farm work, $[1 + (P_Q/P_F)f_1(1 - \alpha)]$ is the implicit wage rate of the wife's home labor and h_2 is the marginal productivity of the wife in home production. The implicit wage rates are not measured or valued in "dollar" terms because the exchange markets are internal. Nonetheless, the "wage rates" are intuitively appealing. For example, the value of an additional hour spent on the farm is measured in terms of what the marginal output from the hour expended in farm labor can produce as input in home production. A similar conclusion is reached about the wage rate implied for home production.

While it would be desirable to estimate a home production function or value function directly in order to obtain estimates of the value of marginal productivity of work at home, unfortunately, output, Z_2, i.e., the wife's home production, is unobserved. Furthermore, it is difficult, if not impossible, particularly in this case, to separate market consumption goods, M, from market goods used as inputs in home production. Instead, we estimate the home production function indirectly by estimating a marginal productivity of work-at-home function. Actually, the marginal (value) productivity function will be estimated through the estimation of a work-at-home function (i.e., a home labor supply function) given that the equilibrium assumption expressed in (7) permits us to use an imputed value for the marginal value product of wives' farm work in place of the household "wage rate." Since the home marginal value product function is the derivative of the value-of-household-production function with respect to the time spent by the wife in the home, integration of this function with respect to T_{Z2} will produce an estimate of the value of a wife's contribution through work at home.

Following Gronau and Heckman[9] we assume that the marginal value product function is of the semi-log form, i.e.,

$$\ln h_2^* = a_1 - a_2 T_{Z2} + a_3 R, \tag{8}$$

where h_2^* = marginal value product of wife's home labor time, i.e., $[1 + (P_Q/P_F)f_1(1 - \alpha)]$ from Equation 7, T_{Z2} = time spent by wife in home production, R = a vector of characteristics of the wife and the household that will affect the wife's marginal productivity, and a_1, a_2, a_3 are parameters to be estimated.

The equation that is most useful for purposes of estimation is a work-at-home function which may be written as

$$T_{Z2} = b_0 - b_1 \ln h_2^* + b_2 R, \tag{9}$$

where the relationships among the parameters in (9) and (8) are given by

$$(b_0/b_1) = a_0$$

$$(1/b_1) = a_1$$

$$(b_2/b_1) = a_2. \tag{10}$$

The "wage rate" h_2^* in Equation 9 is unobserved. To impute a value for the marginal product of work-at-home, it has been assumed the equilibrium condition, i.e., Equation 7 was satisfied and that a reasonable estimate of the value of marginal farm product could be established with reference to the market price of farm output, P_Q, and an estimate of the marginal physical product (MPP) of a wife's farm labor.

This application involved estimating a Cobb-Douglas type agricultural production function with the total market value of farm output as the dependent variable. The labor inputs, T_{Q1} and T_{Q2}, of husbands and wives, respectively, were measured over the entire farming season which is approximately 180 days. An average hourly farm "wage rate" for the wife $f_3^* = (h_1\alpha)f_3$ was then obtained from the output elasticity of women's farm time. Hence for actual estimation purposes we were forced to substitute f_3^* for h_2^* in Equation 9. Equation 9 thus was estimable, obviously subject to some measurement error and other stochastic disturbances. Given that the parameters in (9) could be obtained and the parameters of (8) computed according to the relations presented in (10), an estimate of the total value of a wife's work at home can be computed using the following integration

$$V = \int_0^{T_{Z2}} h(\tau)\, d\tau$$

$$= \int_0^{T_{Z2}} \exp(a_0 - a_1\tau + a_2R)\, d\tau$$

$$= \exp(a_0 + a_2R)[1 - \exp(-a_1T_{Z2})]/a_1, \tag{11}$$

where V represents total value of household production during a year. According to our assumptions and method of imputing wage rates, V will be expressed in "dollar" terms since the home and farm hourly wage rates are expressed in terms of the market price of farm output.

The Estimated Econometric Model

The estimated econometric model involved a stochastic version of the work-at-home function in Equation 9

$$T_{Z2} = b_0 - b_1 \ln f_3^* + b_2 R + u, \tag{12}$$

where u = stochastic disturbance assumed independent of f_3^* and R and identically distributed across all households in the sample. The specific variables characterizing the household and the wife represented by R that we used to estimate the relationship (12) are listed and defined in Table 3.2.

The estimates of the parameters in Equation 8 are obtained from those in (12) by using the relations in (10). Point estimates of value of home production, V, are obtained by computing (11). It is important to note that point estimates of V from (11) are subject to sampling and specification errors, that is, at the very least, estimates of V from (11) are random variables because they are functions of the estimators of a_0, a_1 and a_2.

The uncertainty associated with the use of these point estimates of value for drawing inferences about the economic contribution of children in rural West Africa rests on their randomness. Unfortunately, standard errors are not readily estimated upon which to base additional statistical analyses. A method called the bootstrap for obtaining a nonparametric estimate of the standard error of V was used to solve this difficult problem. Excellent descriptions of the bootstrap method are contained in Bickel and Freedman, Efron and Therneau.[10] This technique also permits us to estimate standard errors for the estimated coefficients in the marginal productivity function, i.e., Equation 8. Thus, we are able to test the significance of the coefficients in the equation and draw some conclusions about the relative importance of the determinants of marginal productivity.

NOTES

1. Drawn from papers jointly authored by R.D. Singh and M.J. Morey. The section on women is taken from the paper published in *Economic Development and Cultural Change* (Chicago: The University of Chicago, 1987).

2. See Susan Tifft, "The Triumphant Spirit of Nairobi," *Time* (August 5, 1985), pp. 38-40. Reporting on the proceedings of the conference marking the end of the United Nations Decade for Women in Nairobi, Kenya, the author commented on the major proceedings, in particular on the question of the economic contributions that women make to national income and welfare and on the neglect of women in the development and planning processes. To quote from Tifft: "In one section of the report, the members urge their countries to put an economic value on the work of women who raise families, keep house and grow crops" (p. 38). Focusing on this rather important aspect, Tifft stated: "Indeed according to a recent survey by Ruth Leger Sivard of World Priorities (Washington, D.C.), the cash value of the unpaid labor of women represents $4 trillion a year, equivalent to a third of the world's gross economic product" (p. 38). The author quoted one of the participating women leaders as saying: "Women are very determined that our work no longer be invisible."

3. Due to data limitation, estimates of the value of work-at-home must be obtained indirectly because the quantities, as well as the prices of goods and services produced within the household (child care, cooking, and product transformation, e.g.) are unobserved. One approach to this problem entails estimating a marginal productivity function for work-at-home. The work of Gronau (1976, 1980) is useful in this effort. However, with Gronau's method the standard errors of the estimates of the coefficients in the marginal productivity function are not automatically estimated, making it difficult to do any sort of statistical inference. This study departs from the previous studies by resorting to the bootstrap technique suggested by Efron (1979, 1982) that allows for obtaining the standard errors for the estimates of the coefficients in the marginal productivity function and of the point estimates of the value of contribution in home production. In this respect, the study provides additional information that would be unobtainable through conventional methods.

4. This has been well focused by the World Bank (the African Review Group) in its "Accelerated Development in

Sub-Sahara: An Agenda for Action" (Washington, D. C.: World Bank, Oct 1981), pp. 74-75.

5. Relating the value of home production with the family farm income is considered extremely relevant for a developing country such as Burkina Faso in the West African region because (a) farm income is the major income earned by households and (b) the major, in fact, the only source of employment outside of the household (for women's labor in particular) is agriculture.

6. Note, that for children, estimates were obtained using broadly the same form(s) of the model, and the estimating equations.

7. We have assumed that two separate household production functions exist, one for the heads, and one for the wives. The rationale or basis for making such an assumption is the traditionally determined male-female roles in home production and in other activities among the household members. The wives more or less exclusively perform home production tasks of cooking, washing, cleaning, food processing (production transformation), and child care, while the males perform tasks such as hunting, fishing, and most of the construction work.

8. All three production functions are assumed to exhibit decreasing marginal productivity.

9. See Gronau, "Home Production--a Forgotten Industry," *Review of Economics and Statistics* 73(1980):97-106.; and James J. Heckman, "Shadow Prices and Labor Supply," *Econometrica* 42(1974):679-94.

10. See P.J. Bickel and E.A. Freedman, "Some Asymptotic Theory for the Bootstrap," *Annals of Statistics* 9(1981): 1196-1217; Efron, "Bootstrap Methods," and "The Jackknife, the Bootstrap, and Other ReSampling Plans"; and T. Therneau, "Variance Reduction Techniques for the Bootstrap," Technical Report No. 200 (Stanford, CA: Stanford University, Department of Statistics, 1983).

REFERENCES

Becker, Gary S. *Human Capital*, 2nd ed. New York: Columbia University Press, 1975.

__________. "A Theory of the Allocation of Time." *Economic Journal* 75(1965):493-517.

Becker, Gary S. and H. Gregg Lewis. "On the Interaction Between Quantity and Quality of Children." *Journal of Political Economy* 81(March/April 1973).

Bickel, P.J. and D.A. Freedman. "Some Asymptotic Theory for the Bootstrap." *Annals of Statistics* 9(1981):1196-1217.

Boserup, E. *Women's Role in Economic Development*. New York, St. Martin Press, 1970.

Cain, Meat T. "The Economic Activities of Children in Bangladesh." *Population and Development Review* 3(1977): 201-8.

Chiswick, Carmel U. "The Value of a Housewife's Time." *Journal of Human Resources* 17(1982):413-25.

Efron, B. "Bootstrap Methods: Another Look at the Jackknife." *Annals of Statistics* 7(1979):1-26.

_________. "The Jackknife, the Bootstrap, and Other Re-Sampling Plans." CBMS-NSF Regional Conference Series in Applied Mathematics, Monograph 38, Society for Industrial and Applied Mathematics, Philadelphia, 1982.

Evenson, R.E. and B.M. Popkin. "Notes on the Laguna Household Study in the Philippines." Mimeographed, Economic Growth Center, Yale University, 1976.

Evenson, R. "On the New Household Economics." *Journal of Agricultural Economics and Development* 6(1970):87-103.

_________. "Time Allocation in Rural Philippine Households." *American Journal of Agricultural Economics* 60(1978):322-30.

Freedman, D.A. "Bootstrapping Regression Models." *The Annals of Statistics* 9(1981):1218-28.

Gronau, R. "The Intrafamily Allocation of Time: The Value of the Housewife's Time." *American Economic Review* 68(1973): 634-51.

_________. "Leisure, Home Production and Work--The Theory of the Allocation of Time Revisited." Mimeographed, National Bureau of Economic Research Working Paper #137, Stanford, California, May, 1976.

_________. "Home Production--A Forgotten Industry." *Review of Economics and Statistics* 73(1980):97-106.

Hawrylyshyn, Otu. "The Value of Household Services: A Survey of Empirical Estimation." *The Review of Income and Wealth* Ser. 22(1976):101-31.

Heckman, James J. "Shadow Prices and Labor Supply." *Econometrica* 42 (1974):679-94.

Hoffman, Wallace E. "Farm Household Production: Demand for Wife's Labor, Capital Services and the Capital-Labor Ration." Agricultural Economic Workshop, the University of Chicago, Paper No. 82:9, 1982.

_________. "Farm and Off-Farm Work Decisions: The Role of Human Capital." *Review of Economics and Statistics* 73(1980):14-23.

Lancaster, K.S. "A New Approach to Consumer Theory." *Journal of Political Economy* 74(1966):132-57.

Lopez, Ignez G.V. "Time Allocation of Low-Income Rural Brazilian Households: A Multiple Job-Holding Model." Unpublished Ph.D. Thesis, Department of Agricultural Economics, Purdue University, August, 1977.

Makhija, Indra. "The Economic Contribution of Children and It's Effects on Fertility and Schooling--A Case Study of Rural India." Workshop in Applications in Economics paper, Department of Economics, University of Chicago, May 3, 1976.

Morey, M.J. and L. Schenk. "Small Sample Behavior of Bootstrapped and Jackknifed Regression Estimators in Misspecified Regression Models." Proceedings of the American Statistical Association, Business Statistics Section, December, 1984.

Nag, Moni, B.N.F. White and Peet R. Creighton. "An Anthropological Approach to the Study of the Economic Value of Children in Java and Nepal." *Current Anthropology* 19(1978):293-306.

Rosenzweig, M.R. and R.E. Evenson. "Fertility, Schooling and the Economic Contribution of Children in Rural India: An Econometric Analysis." *Econometrica* 45(1977):1065-80.

Rosenzweig, Mark R. "Neoclassical Theory and the Optimizing Peasant: An Econometric Analysis of Market Family Labor Supply in a Developing Country." *Quarterly Journal of Economics* (1980):31-55.

Schultz, T.W., ed. "Fertility and Economic Values." *Economics of the Family*. Chicago:The University of Chicago Press, 1974.

__________. "The Value of Ability of Deal with Disequilibria." *Journal of Economic Literature* 13(1975):827-46.

Shortlidge, Robert L. "A Socioeconomic Model of School Attendance in Rural India." Occasional Paper No. 86, Department of Agricultural Economics, Cornell University, Ithaca, NY, January, 1976.

Singh, R.D., E.W. Kehrberg and W.H.M. Morris. "Small Farm Production in Upper Volta: Descriptive and Production Function Analysis." Station Bulletin No. 442, Department of Agricultural Economics, Purdue University, 1984.

Singh, R.D. and M.J. Morey. "The Value of Work-at-Home and Contributions of Wives' Household Service in Polygynous Families: Evidence from an African LDC." *Economic Development and Cultural Change* 35(1987):743-65.

Singh, R.D., G.E. Schuh and E.W. Kehrberg. "Economic Analysis of Fertility Behavior and the Demand for Schooling Among Poor Rural Households in Rural Brazil." Station Bulletin No. 214, Purdue University, Department of Agricultural Economics, West Lafayette, Indiana, 1978.

Therneau, T. "Variance Reduction Techniques for the Bootstrap." Technical Report No. 200, Department of Statistics, Stanford University, 1983.

Tifft, Susan. "The Triumphant Spirit of Nairobi." *Time*, August 5, 1985, 38-40.

Wales, Terence and A.D. Woodland. "Estimation of the Allocation of Time for Work, Leisure and Housework." *Econometrica* 45(1977):115-132.

Walker, K.E. and W.H. Gauger. "The Dollar Value of Household Work." New York State College of Human Ecology, Cornell University, 1973.

World Bank. The World Development Report. Washington, D.C.: World Bank, 1983.

__________. *Accelerated Development in Sub-Sahara: An Agenda for Action*. Washington, D. C.: World Bank, October, 1981, 74-75.

4

Household Migration Decisions

I. INTRODUCTION

Economists have generally used a two-sector model to explain the rural-to-urban migration of labor at the aggregate or regional level, and the labor supply and demand conditions in the urban areas (Todaro 1969; Sjaastad 1961, 1962; Sahota 1968; Berry 1970; Mehmet 1976; and Banerjee 1981). In most of the models, the focus is on the individual, who is assumed to maximize his or her utility or income while deciding about whether and where to migrate. However, in recent years, some studies (for example, Mincer 1978) have drawn attention to family considerations in migration decisions in the context of a nuclear family in which both husband and wife work. The Mincer model assumes that migration decisions involve the movement of the entire (nuclear) family. In such a situation, movement of one individual, or a part of the family, to maximize the individual's utility or income may result in the upsetting or even the breaking down of the family unit. However, this does not hold for most of the African and Asian rural societies in which the extended family system is still the rule and the household acquires a multicentered character. The migratory behavior of individual members reflect, in such settings, family decisions. Often, even after the household member migrates, he generally maintains an essential link with the household through: (a) frequent visits to the home village with "gifts" of all sorts for the parents and other relatives, and (b) money remittances (Banerjee 1981; Singh 1978, 1981). Such a unique relationship adds another dimension to the migration process not found in the more industrialized societies. Therefore, the household characteristics and constraints that may shape family migratory decisions cannot be ignored, although the fact is that most past studies have, for the lack of household level data, ignored such characteristics and relied on aggregate regional data.

The purpose of this study is to identify, in the setting of rural Western sub-Saharan Africa, the best set of economic variables which operate at the household level to determine the probability that a rural household will allow part of its labor force to emigrate. It is recognized that there are important sociological and cultural factors which shape the typical rural household's interaction with the external sector. Although it is difficult to test these relationships empirically, it is believed that these sociological and cultural factors mostly work to preserve the cohesiveness of the family unit and the traditional way of life, and, hence, they may tend to have a negative impact on out-migration. Ultimately, however, economic forces induce the household to expand the geographical scope of the markets in which its labor force is deployed. In the rural sub-Saharan setting, the scarcity of good quality farm land seems to be a major factor in rural-urban migration. This study hypothesizes that the scarcity of farmland is the major "push" factor in the migratory behavior of farm population in rural Africa.

The land-migration relationship for a country in the sub-Saharan African setting needs to be viewed from two aspects which have important bearing on household's migration decisions. First, the farming system in the region is highly traditional and labor-intensive. In most of the sub-Saharan African countries, but unlike the Asian and the Latin American countries, there is no class of landless agricultural labor in the rural areas. Every family in the village has some land area to farm, although every family does not operate the same size of farm. There is practically no hiring of paid labor on the farm from outside, as traditionally no such market exists in rural areas. Most of the farm labor supply comes from the households themselves. Second, there are very limited, or almost no, off-farm employment opportunities for rural labor in and around the villages which could deter out-migration. In a setting such as this, the demand for labor is determined largely by the size of the family farm. The combined effect of the existing farming systems and the lack of off-farm avenues of employment is an extremely low marginal productivity of farm labor, in particular the marginal productivity of family male labor (Ram and Singh 1984). Therefore, it is only reasonable to expect that families with smaller farms and lower per adult farm land area will seek to maximize their utility and income by sending out a part of their adult labor force in search of more remunerative work. It is no accident then that the "Mossi" region, which is also the most populous region, has, over time, witnessed one of the highest rates of out-migration (Census Reports, Burkina Faso).

Although farm size is considered a dominant factor in rural migration, the effects of some other economic variables are also studied. These are: distance to the main urban centers in the area (the urban centers in the area serve as the first stop for the migrant and also provide some employment), access to information about economic conditions and employment prospects outside of the village, and the amount of schooling acquired by household members. Specifically, the study attempts to evaluate the effects on the household's migration decisions of: (i) the family farm size in relation to the household's active labor supply, (ii) information supply, represented by the presence of the household's previous migrants in the destination areas, and by the location of the villages with respect to urban centers; and (iii) the level of schooling.

Although several studies on the subject are available for most advanced countries and for a limited number of the developing countries, there is a paucity of empirical studies on household migration for most of the African developing countries. This is due primarily to the lack of cross sectional household data. The availability of a unique set of household data from Burkina Faso has enabled this study to analyze the household migratory decisions. Furthermore, since the analysis is conducted at the disaggregated (household) level, the findings will help broaden our understanding of the migration decision-making process at the level of the household, and the economic basis for the decisions which are eventually made.

II. THE HYPOTHESIS AND THE MODEL

The household's decisions with respect to labor migration can be broken down into two parts: first, the household has to decide whether or not to participate in the outside economy as a means of enhancing its income earning opportunities; and, second, it has to decide on the intensity of participation in the outside labor market. It can be hypothesized that the observed variations across households with respect to these two are attributable to variations in the household's labor needs, and to the direct cost involved in sending a member of the household to the urban centers and other destination points. The labor needs and the cost variables are measured indirectly through proxies in the model. The household's labor needs are assumed to be directly related to the size of the farm operated by the household. The cost to the household of sending out a migrant is assumed to be related to the availability of information and the distance of the village to the main urban centers in the region. It is assumed that the households in villages with prior family migrants, or those located closest to the urban-industrial centers, must experience lower migration costs, all other things equal. Households with prior migrants have been compared to early "adopters" of innovations in some of the developing countries' agriculture, such as the use of thigh yielding varieties of seeds and fertilizers etc. (Lipton 1976; Rogers 1968).

Two alternative formulations of the migration model have been used to test the hypothesis that difficulties in obtaining an adequate amount of farmland is one of the major determinants of the probability that a household will send out a migrant. In the first formulation, Y is assumed to represent a dichotomous variable which takes on the value of 1 if the household sent out a migrant during the survey period (1977 to 1980), and 0 otherwise. For the ith household we have:

$$P_i = P(Y_i = 1) = F(X_iB), \tag{1}$$

where X_i denotes a vector of independent variables, and B denotes a vector of unknown parameters. If F is of the form

$$F(X_iB) = X_iB, \tag{2}$$

then (1) is the so-called linear probability model (Pindyck and Rubinfield 1976; Amemiya 1981). For the ith household, the equation implies that whether or not the household chooses to participate in the outside economy directly by sending out a migrant depends on the X vector assuming some threshold values, X*. In the above model, the X vector essentially represents the variables that determine the differential between labor requirements and availability in the household. The linear probability model provides a simple and straightforward means of testing models with dummy dependent variables.

The other alternative formulation of the model to test the hypothesis on migration focuses on the intensity of participation of the labor force in the outside market. The form of the model is specified in the following section that discusses the results of the regression analyses.

III. THE EMPIRICAL RESULTS

The major findings of the study are reported in this section in two parts: the first part provides a brief descriptive account of the pattern of migration, the characteristics of the migrants including the linkages these migrants maintain with their families; while the second part presents the results of the statistical analysis of the household's migratory behavior. The major focus of the discussion is on the second part, however.

Migration and the Migrants: A Brief Descriptive Account

The survey data indicate that about 66 percent of all farm households in the sample in the regions studied had migrant members living outside the village. Of the total number of migrants, 70 percent had migrated to the Ivory Coast, 25 percent to the country's capital city of Ouagadougou, I percent to Ghana, and the rest to other cities of the country. There were, on the average, about two males and one female migrants per household in the sample. Generally, the first to move out of the farm is usually the adult male member, later to be followed by the spouse(s) and children. Overall, the migrants accounted for over 20 percent of the total population in the sample (Table 4.1).

Most of the migrants are adult males and they belong to the relatively small farm households. For example, over 70 percent of all the migrants are from households which have less than 5 hectares, or approximately 12.5 acres, of land area farmed (Table 4.2). As one would expect, it is the relatively low income households with small farms and poor economic bases in the village that would send out adult male workers in search of better employment opportunities. The data in the table support this pattern, particularly with respect to the land base.

The other point that needs to be focused is the contributions of the migrants to their households. According to this study's estimates, the households with a part of their labor force working outside the village received annually, through money remittances, $30.00, on average. Although this is a small amount in an absolute term, this is a little over 4 percent of the family's total annual income. Also, note that this cash

TABLE 4.1
Distribution of Migrants by Age: Farm Households, Burkina Faso, 1980

Age	All Migrants		Males		Females	
	Freq.	Percent	Freq.	Percent	Freq.	Percent
0 to 6	28	18.4	10	10.0	19	36.5
7 to 14	13	8.6	8	8.0	5	9.6
15 to 19	16	10.5	8	8.0	8	15.4
20 to 24	30	19.7	23	23.0	7	13.5
25 to 29	26	17.1	19	19.0	7	13.5
30 to 34	16	10.5	13	13.0	3	5.8
35 to 39	9	5.9	9	9.0	0	0.0
40 to 44	4	2.6	3	3.0	1	1.9
45 to 49	1	0.7	1	1.0	0	0.0
50 to 54	4	2.6	3	3.0	1	1.9
55 to 59	2	1.3	2	2.0	0	0.0
60 or more	2	1.3	1	1.0	1	1.9
	152	100.0[a]	100	100.0	52	100.0

[a]Totals may not add up to 100 due to rounding.

Source: Farming Systems Research, Sample Survey 1979-80.

contribution of the migrants is in addition to the "gifts" received in kind (such as clothes, bicycles, grains, and sometimes even farm equipment and other inputs) by the families of the migrants.

Household's Migration Behavior: Results of Regression Analyses

The model that has been used to estimate the migration relationships is of the following form:

$$MG_i = b_1 + b_2 LANDN_i + b_3 DST_i + b_4 EDUC_i + b_5 MGB_i + b_6 AVAG + e_i \qquad (3)$$

TABLE 4.2
Distribution of Sample Households and Migrants by Farm Size, Burkina Faso, 1980

Farm Size[a]	All Migrants		Adult Migrants		Households	
	Number	Percent	Number	Percent	Number	Percent
Under 3.0	29	25.9	19	25.7	13	22.4
3.0 to 6.0	51	45.5	33	44.6	30	51.7
6.0 to 9.0	25	22.3	17	23.0	11	19.0
9.0 to 12.0	0	0.0	0	0.0	0	0.0
12.0 to 15.0	2	1.8	1	1.4	2	3.4
15.0 to 18.0	0	0.0	0	0.0	0	0.0
18.0 to 21.0	0	0.0	0	0.0	0	0.0
21.0 or more	5	4.5	4	5.4	1	1.7
	112	100.0[b]	74	100.0	58	100.0

[a]In hectares, where 1 hectare = 2.47 acres.
[b]Totals may not add up to 100 due to rounding.

Source: Farming Systems Research, Sample Survey, 1979-80.

where MG_i = 1 if the ith household sent out a migrant, and 0 otherwise; $LANDN_i$ = land area farmed per adult by the ith household; DST_i = 1 if the ith household is located in the village which is closest to the capital city of Ouagadougou (which is the country's largest city), and 0 otherwise; $EDUC_i$ = the average years of schooling of the male members of the ith household (all females in the sample were illiterate); MGB_i = 1 if the ith household has prior migrant member in the urban center (i.e., the migrant's destination point), 0 otherwise; AVAG = the average age of the adult members in the households,[1] and e_i is the error term. The mean values and coefficients of variations of the variables of the migration model are provided in Table 4.3.

The coefficients of the variables included in the model were estimated using the ordinary least-squares (OLS). As evident from the results presented in Table 4.4, the coefficients on the variables appear stable across alternate specifications of the estimating model. The adjusted R^2 indicates that 37 to 40 percent of the cross-sectional variance in the model is explained

TABLE 4.3
Mean Values and Coefficients of Variation of Selected Variables

Variables	All Households		Non-Migrant		Migrant	
	Mean	Coef.(%)	Mean	Coef.(%)	Mean	Coef.(%)
Number of Household Members	14.897	56	12.440	71	16.758	46
Number of adult members (15 and above)	8.276	53	6.440	64	9.667	43
Number of migrants (15 years and above)	1.931	113	--	--	3.364	56
Number of migrants (15 and above) who left after 1976 (MG15 76)	1.276	113	--	--	2.242	54
Size of household farm in hectares (LAND)	5.218	76	4.966	72	5.410	79
Amount of farmland per adult member of household (LANDN)	.697	55	.868	50	.568	51
Average age of household members (AVAG)	22.974	26	22.560	28	23.288	24
Average age of adult household member (AVAG 15)	34.546	17	34.544	17	34.548	16
Average years of schooling of adult male members (YE7)	1.554	144	0.808	161	2.120	123
Total years of schooling of household (adult male) members (YE4)	5.638	220	2.240	145	8.212	193

TABLE 4.4
Estimated Coefficients of the Migration Model:[a]
Farm Households, Burkina Faso, 1980

Independent Variables	Equations: (1)	(2)	(3)	(4)
Farmland per active household member (LANDN)	-.4922 (3.420)[b]	-.4591 (-2.803)	-.4899 (-3.375)	-.4881 (-3.222)
Presence of prior migrants at destination points (MGB76)	.5057 (4.33)	.4875 (4.139)	.5168 (4.319)	.5313 (4.368)
Distance (location) of villages from main urban centers (DST)	.1647 (1.410)	.1603 (1.292)	.1588 (1.34)	.1670 (1.41)
Total schooling years of male members (YE4)	--	--	.0024 (.540)	--
Average schooling years of male members (YE7)	--	.0128 (.499)	--	.0024 (.089)
Average age of household members 15 years and above (AVAG15)	--	--	.0172 (1.30)	.0168 (1.198)
Average age of household members 15 years and over squared ($AVAG15^2$)	.0024 (.560)	--	.0000 (-.499)	.0000 (-.467)
Average age of household members (AVAG)	.0124 (1.36)	-.0022 (-.034)	--	--
Average age of household members squared ($AVAG^2$)	--	.0001 (.120)	--	--
Constant	.2372 (.69)	.6154 (.719)	.0991 (.225)	.1117 (.248)
$\bar{R}^2$	.40	.37	.39	.39
F	5.58	6.57	7.09	7.00

[a]The dependent variable is MG_1. MG_1 = 1 if the household sent out a migrant, and 0 otherwise.
[b]t-statistics in parentheses.

by the independent variables included in the estimating model. The F-value is significant at the one percent level implying that the systematic variation is considerably larger than should be expected by chance.

The coefficient of the land variable, LANDN, the per adult member land area farmed, appears, as expected, consistently negative and statistically significant at less than the one percent level. This implies that the scarcity of farm land is an important causal factor in explaining the migratory behavior of farm households. In dominantly traditional, labor-intensive farming systems, households with relatively larger-sized farms and greater amount of land to cultivate per active person would tend to have greater demand for labor services on the farm. This means that as the size of the farm increases, other things constant, there will be less tendency for household members to move out. The result supports the land-push hypothesis in family migration decisions.

The result pertaining to the land-migration connection is important, for according to some writers (Songre 1973; Fleury 1979, for example), there is no shortage of land in the region, and, furthermore, the short-fall in farm production is attributed to labor shortages caused by out-migration of labor from rural areas. The present evidence does not seem to support this view. The results of this study indicate that in a land-scarce situation of the type found in the sub-Saharan region, it would seem economically rational for a part of the household's labor force to seek out migration with the objective of maximizing its utility or income. Furthermore, money remittances and other "gifts" in kind received from outside provide sources of additional income to migrant families in the villages.[2] Also note that the estimated per hectare crop yields and net farm income for the sample households in the study regions of Burkina Faso show that the per hectare yields and net farm income were 15 to 20 percent higher for the migrant households than the non-migrant households.

The coefficient of the prior migratory experience variable, MGB76, is positive and statistically significant at less than the one percent level (Table 4.4). The powerful impact of the prior experience variable on migration may be assumed to represent the effects of: (a) the easy and almost free inflow of information about outside jobs to farm households which have prior migrants located in the urban industrial centers; and (b) the reduced risks and uncertainties, for the prospective migrants, involved in moving, waiting and looking for jobs in strange areas. The result of the combined effects of (a) and

(b) is to lower the costs involved in migration. The supply of information from the previous migrants through visits and letters is indeed important to farm households whose members are mostly illiterate and to whom other (formal) sources of information are almost inaccessible. Similarly, the presence of such members in cities in or out of the country[3] means, for the prospective migrants, that there is a place to live and food to share with relatives (the prior migrants) who generally take care of such members and help them in finding jobs.

The coefficient of the education variable, EDUC, the average school years of the household's male members, has the predicted positive sign, but it is not statistically significant. The schooling of the household head did not show any effect on migration, and, therefore, this variable was dropped from the estimating model. The weak schooling-migration relation appears reasonable to expect in view of the extremely low level of schooling among the rural population studied, with little or no variations across households. Schooling is mostly at the lower primary level, in many cases involving a year or two spent in learning Arabic. The coefficient of the average age variable, AVAG, appears positive, but statistically weak. The coefficient of the distance dummy variable, DST, also has the expected positive sign, but again statistically not significant.

Another Specification of the Migration Model and the Regression Results

The migration behavior of households has been analyzed also through an alternate specification of the estimating model 4 (Equation 4), in which the dependent migration variable is measured by the proportion of the household's adult labor force that migrates out to urban-industrial centers, whether within or outside of the country. This alternate measure of migration indicates the intensity of the participation of the farm labor force in the external labor market. However, the explanatory variables of model 4 remain the same as in the previous model (Equation 3).

$$MG_{2i} = c_1 + c_2LANDN_i + c_3DST_i + c_4EDUC_i + c_5MGB_i + c_6AVAG + U_i \qquad (4)$$

where MG_2 = ratio of the adult migrants to the total adult members of the household, and U_i = the error term (the other variables are already defined).

The results of the estimated coefficients (Table 4.5) strongly provide further evidence of a negative impact on labor retention on the farm resulting from too little land. As indicated by the negative and statistically significant coefficient of the land variable, the households with smaller landholdings per adult member in the household tend to be the ones with larger proportions of their adult members migrating out of the region. The coefficient of the prior migration variable (MGB) also appears positive and statistically significant at the five percent level or lower. As indicated by the results in Table 4.5, the average age variable (AVAG) shows a much stronger (positive) effect on the household's migration intensity than it did in the previous case (Table 4.4). The other variables of the alternate model have coefficients with the expected signs, although they appear statistically weak. Overall, the results estimated through model 4 (Table 4.5) are similar to those obtained through model 3 (Table 4.4). The signs as well as the levels of statistical significance of the variables of the migration model stay stable and consistent across the several alternate specifications of the model.

IV. SUMMARY AND CONCLUSIONS

The major findings of the study may now be summed up as follows. First, the scarcity of good quality farm land and, hence, the low demand for labor services on the family farm, provide adult members of rural households the incentive to seek out-migration. The majority of the migrants are adult male members of the household. The result of a negative and statistically strong coefficient of farm size shows that the land-push hypothesis does, in fact, hold. Second, the result also highlights the role of information supply to households through prior migrants. This further enhances the mobility of farm workers. Both the prior migratory experience variable and the location variable exercise positive effects on migration. The estimated coefficient for the prior migration variable further implies that chain migration favors the family that is an early "adopter" of migration. The early "adopters" of migration can more readily acquire, process, and use information sent by their relatives, and thereby reduce the costs (also some of the risks) of migration. Therefore, it is no surprise that the

TABLE 4.5
Estimated Coefficients of the Family Migration Equations:[a] Farm Households, Burkina Faso, 1980

Independent Variables	Equations: (1)	(2)	(3)	(4)
Farm land per active person (LANDN)	-.1594 (2.94)	-.1286 (-2.055)	-.1569 (-2.923)	-.1551 (-2.78)
Presence of prior migrants at destination points (MGB76)	.0831 (1.89)	.0646 (1.435)	.0949 (2.145)	.0948 (2.114)
Distance (location) of villages (DST)	.0331 (.75)	.0190 (.400)	.0268 (.610)	.0270 (.617)
Total schooling of male members (YE4)	--	--	.0002 (.136)	--
Average schooling years of male members (YE7)	.0003 (.19)	.0087 (.885)	--	.0012 (.126)
Average age of household members 15 years and above (AVAG15)	.0072 (2.10)	--	.0213 (2.507)	.0120 (2.327)
Average age of household members 15 years and above squared ($AVAG15^2$)	--	--	.0000 (-1.428)	.0000 (-1.328)
Average age of household members (AVAG)	--	.0081 (.332)	--	--
Average age of household members squared ($AVAG^2$)	--	-.0001 (-.243)	--	--
Constant	-.0375 (.29)	.0674 (.206)	-.1837 (-1.129)	-.1791 (-1.079)
$\bar{R}^2$	.18	.12	.20	.20
F	3.54	2.24	3.35	3.35

[a]The dependent variable is MG_2, the percentage of the household's adult members who have migrated.
[b]t-statistics in parentheses.

presence of prior migrants provides a strong stimulus to small farm households to send out members to participate in the external labor market.

NOTES

1. The average age variable was specified in two alternate ways: one for all the household members and the other for only the adult members 15 years old and above.

2. According to some estimates (the World Bank, the Government of Burkina Faso), the recorded migrants' remittances from abroad in 1981 to Burkina Faso amounted to nearly one-half of the official exports of goods and services (the unrecorded remittances may be even higher). About one-half of the gross annual migrants return to Burkina Faso every year, and with them come some very much needed skills. Courel and Pool (1976) estimated the average gain per migrant from Burkina Faso from each migration (1956-60) as follows:

Gain (US $) Per Migrant

Country of employment	Cash ($)	Kind ($)	Total ($)
Ivory Coast	38	19	57
Ghana	24	17	41
All other countries	182	49	231
All countries	39	20	59

These data further add to the evidence that the households which participate in the external labor force and interact with the external market do benefit through cash remittances and other transfers from the migrants.

3. It is indeed interesting to point out that of all the migrants, as many as 5 percent had at one point or the other, migrated out of the country (of these 70 percent were in Ivory Coast and 1.5 percent in Ghana), while the remaining 29.5 percent migrated to urban centers within the country (of the

latter, Ouagadougou, the capital city accounted for 24 percent, and the other cities, for the remaining).

REFERENCES

AdePoju, Aderanti. "Rural-urban Socio-Economic Links: The Example of Migrants in South-West Nigeria." In *Modern Migrations in Western Africa*, ed. by Amin Samir, pp. 127-137. London: Oxford University Press, 1974.

Amemiya, Takeshi. "Quantitative Response Models: A Survey." *The Journal of Economic Literature*, 19(1981): 1483-1546.

Banerjee, Biswajeet. Rural-Urban Migration and Family Considerations in Migration Behavior in India. *Oxford Bulletin of Economics and Statistics* 43(1981):321-353.

Beals, R.E., M.B. Levy, and L.N. Moses. "Rationality and Migration in Ghana." *Review of Economics and Statistics* 49(1967):400-86.

Berry, Sara S. "Economic Development with Surplus Labour: Further Complications Suggested by Contemporary African Experience." *Oxford Economic Papers* 22(1970):275-82.

Courel, Andre and D. Ian Pool. "Upper Volta." In *Population Growth and Socio-Economic Change in West Africa*, ed. by Caldwell, J.C., pp. 736-54. New York: Columbia University Press, 1975.

Elkan, Walter. "Migrant Labor in Africa: An Economist's Approach." *American Economic Review*, Papers and Proceedings 49(1959):188-97.

Fleury, Jean-Marc. "Upper Volta Population on the Move." *The International Development Research Center, Quebec, Canada, Reports* 8(March, 1979):6-7.

Gugler, Josef. "On the Theory of Rural-Urban Migration: The Case of Sub-Saharan Africa." In *Migration*, ed. by A. Jackson, pp. 134-55. Cambridge University Press, 1969.

Lipton, Michael. "Migration from Rural Areas of Poor Countries: The Impact of Rural Productivity and Income Distribution." Research Workshop on Rural-Urban Labor Market Interactions, Development Economics Department, IBRD, Washington, D. C, 1976.

Mehmet Ozay. "A Note on Unemployment and Labor Migration in Less Developed Countries: A Diagrammatic Illustration." *American Journal of Agricultural Economics* 58(1976):351-54.

Mincer, Jacob. "Family Migration Decisions." *Journal of Political Economy* 86(1978):749-73.

Pindyck, Robert S. and Daniel L. Rubinfield. *Econometric Models and Economic Forecasts*. New York: McGraw-Hill, 1976.

Ram, R. and Ram Singh. "Farm Households in Rural Upper Volta: Some Evidence on Allocative and Direct Returns to Schooling and Male-Female Productivity Differentials." Working Paper, 1984

Rogers, E. and L. Svenning. *Modernisation Among Peasants*. New York: Holt, Rinchart and Winston, 1968.

Sahota, G.S. "An Economic Analysis of Internal Migration in Brazil." *Journal of Political Economy* 76(1968):218-45.

Singh, Ram D. "Labour Migration and its Impact on Employment and Income in a Small Farm Economy." *International Labour Review*, International Labour Organization 116(1977): 331-41.

_________. "Small Farm Production Systems in West Africa and their Relevance to Research and Development: Lessons from Upper Volta." The University of Chicago, Agricultural Economics (Department of Economics) Workshop Paper No. 81:20, May 28, 1981.

Sjaastad, L.A. "Income and Migration in the United States." Unpublished Thesis, The University of Chicago, 1961.

_________. "The Costs and Returns of Human Migration." *Journal of Political Economy* 70(1962):80-93.

Songre, Ambroise. "Mass Emigration from Upper Volta: The Facts and Implications." *International Labour Review* 108(1973):209-25.

Todaro, M.P. "A Model of Labor Migration and Urban Unemployment in Less Developed Countries." *American Economic Review* 59(1969):138-48.

PART THREE

Economics of Farming Systems

5

The Economics of Small Farms, the Traditional Farming System and Schooling

I. INTRODUCTION

The majority of farmers in West Africa have small land holdings and under low yield conditions produce subsistence crops to satisfy family needs. Cropped land per capita ranges from as low as 0.1 hectare in Cape Verde to 3.2 hectares in Niger. Operational holdings per household in general are small with about one hectare (2.47 acres) of land per person in the household, and in some cases in some years family farms do not produce enough to meet the household's needs. The main cereals produced and consumed by small-farm families are millet, sorghum, and corn which together account for over 70 percent of the total area devoted to cereals. Agro-economic indicators for countries in West Africa and a select group of countries in semi-arid regions of Africa are presented in Tables 5.1 and 5.2. These countries have drawn considerable attention lately from the international community, donor countries and international organizations alike.

All countries in West Africa are net importers of cereals (Table 5.1) and most of these countries are chronically food deficit countries with frequent droughts. Extremely low farm productivity is reflected in low yields which in most cases range from 300 kg (kilograms; 1 kg = 2.20 lb) to 700 kg of grain per hectare of land (270 pounds to 625 pounds per acre). Perhaps these are the lowest yields in the world. Poor soils, unfavorable and often unpredictable climatic conditions, lack of improved technologies for rain-fed cereal crops, disincentives created by government marketing and pricing policies are factors in the slow growth in farm productivity in almost all of these countries.

TABLE 5.1
Population, Income and Agricultural Economic Indicators for Some Selected Semi-Arid Countries of Africa

SAFGRAD Countries (Africa)	Population Mid-1982 (millions)	GNP Per Capita in US$, 1982	Cereal Yield[a] (ton/ha)	Cereal Output[a] (million tons)	Cereal Cons./Caput[a] (kg/year)	Fertilizer Cons. NPK (kg/ha) 1976	Tractor Density (no./1,000 ha) 1976	Percent of Cereal Imported[a]
Ivory Coast	8.9	950	0.9	0.7	119	5	0.3	20
Zambia	6.0	640	0.9	1.2	252	13	0.8	10
Senegal	6.0	490	0.6	0.7	210	16	0.1	28
Nigeria	90.6	860	0.6	8.4	145	5	0.3	10
Botswana	0.9	494	0.6	0.1	186	2	1.4	32
Ghana	12.2	360	0.7	0.6	73	8	1.2	21
Cameroon	9.3	890	0.9	0.9	128	2	0.0	8
Sudan	20.2	440	0.6	2.6	145	14	1.2	2
Togo	1.8	340	0.8	0.3	131	1	0.1	6
Kenya	18.1	390	1.3	2.2	160	25	2.8	Ex[b]
Mauritania	1.6	470	0.3	0.0	135	1	n.a.	69
Central Afr. Rep.	2.4	310	0.5	0.1	57	0	0.0	10
Guinea	5.7	310	0.7	0.7	177	0	0.0	7
Sierra Leone	3.2	390	1.4	0.6	206	0	0.1	6
Benin	3.7	310	0.7	0.3	110	1	0.0	11
Gambia	0.6	370	0.8	0.1	198	10	0.3	28
Tanzania	19.8	280	0.8	1.5	113	5	1.2	13
Niger	5.9	310	0.4	1.2	271	0	0.0	3
Cape Verde	0.3	340	0.5	0.0	131	4	0.8	90
Guinea Bissau	0.8	190	1.0	0.1	223	1	0.0	25
Chad	4.6	80	0.5	0.6	145	1	0.0	3
Somalia	4.5	290	0.6	0.2	110	n.a.	1.2	34
Ethiopia	32.9	140	1.0	4.9	174	2	0.3	1
Burkina Faso	6.5	210	0.5	1.1	186	1	0.0	2
Mali	7.1	180	0.7	1.1	203	1	0.1	6

[a]For the years 1975-1977.
[b]Codes: n.a. = not available, Ex = net exporter.
Source: World Bank, FAO, International Agricultural Development Service, and government publications, 1975-1980.

TABLE 5.2
Distribution of Arable Land Area Under Major Crops in Selected Semi-Arid African Countries, 1975-1977

SAFGRAD Countries (Africa)	Arable Land as % Total Land	Cereal Area (million ha)	Cereal Area as % Arable Area	Maize as % Cereal Area	Sorghum as % Cereal Area	Millet as % Cereal Area	Rice as % Cereal Area	Wheat as % Cereal Area
Ivory Coast	--	0.7	--	--	--	--	--	--
Zambia	6.6[a]	1.3	11.9	--	--	--	--	--
Senegal	11.7[a]	1.1	--	5.0	--	81.2	13.8	--
Nigeria	--	13.0	50.1[a]	16.5	42.8	34.3	16.4	1.0
Botswana	--	0.2	--	--	--	--	--	--
Ghana	4.4	0.8	88.5[a]	46.4	19.6	14.3	9.6	--
Cameroon	14.2	0.8	12.6	49.1	--	19.1	1.8	--
Sudan	2.8	4.1	53.6	0.6	70.4	17.9	--	11.0
Togo	--	0.3	--	--	--	--	--	--
Kenya	2.7	1.7	11.3	73.5	--	16.5	1.6	6.8
Mauritania	1.0	0.2	16.7	9.1	--	90.9	--	--
Central Afr. Rep.	--	0.2	--	--	--	--	--	--
Guinea	16.9[a]	1.0	24.7[a]	27.7	--	12.2	60.1	--
Sierra Leone	--	0.4	--	--	--	--	--	--
Benin	--	0.4	--	79.1	17.8	3.1	--	--
Gambia	--	0.1	--	--	--	42.9	57.1	--
Tanzania	11.8	2.0	7.5	33.3	--	57.0	9.2	0.8
Niger	11.8[a]	2.9	18.6[a]	0.3	28.7	65.4	5.6	--
Cape Verde	--	0.0	--	--	--	--	--	--
Guinea Bissau	--	0.1	--	--	--	--	--	--
Chad	5.4[a]	1.1	13.0	1.2	--	91.4	6.4	1.0
Somalia	1.6	0.4	54.5	--	--	--	--	--
Ethiopia	10.6	5.1	45.5	25.1	17.6	3.9	--	16.5
Burkina Faso	19.4	2.2	41.9	7.1	58.8	30.8	3.3	--
Mali	9.4[a]	1.5	12.6[a]	6.9	--	69.8	23.3	--

[a]Includes permanent crops.

Source: World Bank, FAO, International Agricultural Development Service, and government publications.

Rainfed crops have by and large lagged far behind irrigated crops (rice in particular) in development of technologies for more economical and higher yields. No technological breakthrough is in sight for cereal crops, particularly with regard to varietal improvements. Those new varieties of sorghum, millet and maize that have been or are being developed and/or tried by plant breeders and agronomists for the low rainfall and high risk regions of Africa have not yet been demonstrated to be superior to current local varieties. It is against this background that the study of the existing farm production systems practiced by the Voltaic farmers (in Burkina Faso) and the constraints confronting them is presented in this chapter which may help appreciate the problems of low productivity and low farm income in African countries, and eventually aid the search for solutions.

The discussion in the following sections of the chapter focuses on: an overall description of agriculture in Burkina Faso in section II; the characteristics of the existing farming systems, crop patterns and crop yields in the three study regions and on the sample farms in section III; household grain production, marketed surplus and prices in section IV; input use, credit constraint and the demand for production credit in section V; production function for major crops estimated through regression analyses in section VI; the estimated marginal value product of production inputs in section VII; and the estimates of the household (aggregate) level production function, the male-female productivity differential and the economic effects of education on farm production in section VIII.

II. AGRICULTURE IN BURKINA FASO: AN OVERALL VIEW

Burkina Faso is landlocked by Mali on the North and West, Ivory Coast, Ghana, and Togo on the South, and Benin and Niger on the East. The land area is 274, 200 km^2 (106,500 mi^2) with an estimated population of 6.5 million (mid-1982). Eighty percent of the country's total population is engaged in agriculture. The latest population growth rate estimate is 2.6 percent per annum.

Most of Burkina Faso lies in the Sudan vegetative zone. Annual rainfall varies from 500 mm (northeast) to 1500 mm (southwest). More than 100 mm of rainfall per month occurs in 4-5 months of the year with the maximum occurring in August. Most of the soils are classified as ferruginous tropical. Sands covered by laterite crusts are extensive in the northeast, southwest, and central regions. Soils of southern and eastern regions were developed from granite, gneisses, and schists. Soils are generally lacking in fertility, and in scanty rainfall areas may be very hard to plow.

Among the semi-arid African countries, Burkina Faso, Mali, Ethiopia, Somalia and Chad, rank lowest in terms of per capita income with ranges between 80 and 290 US dollars, and literacy rate which is not more than 5 to 10 percent of the total adult population. By most major economic and agricultural indicators (Tables 5.1 and 5.2) Burkina Faso can be rated as one of the least developed among the low income countries.

Arable land, however, constitutes only 19.4 percent of the total available land. According to the 1975-1977 data, Burkina Faso has had 2.2 million hectares of land under cereals which accounts for about 42 percent of the country's total arable land. Cropped land per capita amounts to 0.9 hectares. The major cereal crops produced in the country are sorghum, millet and maize. The area percentages devoted to major cereal and other crops per farm estimated by the Directorate of Agricultural Services (Burkina Faso 1974-1975) are as follows: sorghum, 36 percent; millet, 29 percent; maize, 5 percent; rice, 3 percent; cowpeas, 3 percent; peanut, 7 percent; and cotton, 7 percent.

Cereal yields for the country average around 500 kg per hectare. Production of cereals varies from 170 kg to 186 kg per capita per annum. Of the total cereal consumption, imports accounted for two percent of the country's total consumption during 1975-1977 (7 percent during 1973-1974). The data presented in Tables 5.3 through 5.6 demonstrate the gap between estimated requirements and production. The question is how to augment supply to meet the growing need for food by an increasing number of people. With the current average yield level of 500 kg per hectare under cereal crops, the task at hand is undoubtedly difficult. Assuming a 2.6 percent rate of population growth, total food production, for example, will have to increase almost 30 percent by 1990 in order to maintain the current per capita consumption level without a greater proportion of imports. Various questions arise with respect to the problem of increasing production levels. For example, can and should extensive farming be promoted if additional land is available for cereal production or should intensive cultivation practices be encouraged, if the necessary inputs are or can be made available to farmers? There are no obvious answers to these questions.

TABLE 5.3
Production Levels[a] and Trends in Burkina Faso

Years	Sorghum	Millet	Maize	Paddy Rice	Cowpea (dry)	Peanut (in shell)	Pulses Beans
1961-65	514	300	100	34	71	58	--
1970	563	378	55	34	65	68	--
1971	493	277	66	37	60	66	--
1972	512	266	59	30	60	60	--
1973	481	253	58	32	50	63	--
1974	400	220	50	25	55	40	--
1975	738[b]	383[b]	84[b]	40[b]	--	90	180
1976	534	347	60	36	--	72	180
1977	634	354	73	37	--	57	165
1978	610[c]	406[c]	100[c]	28[c]	--	70[c]	180[d]
1979	610	431	100[c]	30[c]	--	75[d]	190[d]
1980	559[d]	330	100[d]	30[d]	--	77[d]	190[d]
1981	750	400	--	--	--	77[d]	--

[a]All production levels in units of 1,000 metric tons.
[b]FAO--official statistic.
[c]FAO--unofficial estimate.
[d]FAO--estimate.

Source: Ministry Rural Development, Government of Burkina Faso and FAO yearbook of production except as footnoted.

The farmer is a principal actor in the production-consumption process. He is influenced by a number of factors over some of which, the exogenous ones, he has no control, and which may seriously constrain his production efforts. An important concern for farming systems research is to find appropriate technological innovations that raise the productivity of agriculture, and for public policy an important concern is the diffusion of such innovations. Innovation may take various forms, for example, improved seeds that are disease resistant and high yielding; use of chemical fertilizers, insecticides and pesticides; introduction of better management practices; and substitution of capital equipment, machinery, and animal traction for labor.

There could be some attractive propositions with regard to new crop varieties. For example, ICRISAT's[1] new sorghum

TABLE 5.4
Estimated Levels of Cereal Consumption[a] in Burkina Faso

Crop	1970[b]	1980	1985	1990
Millet/Sorghum	130	131	131	130
Maize	11	11	12	12
Rice	4	4.5	5	5
Wheat	4	5	5	5.5
Cowpea	20	21	21.5	22
Peanut	6	6	6	6

[a]In kg/capita/year. Estimates for 1970 are based on actual consumption and others upon FAO projections considering elasticity of demand. Taken from International Fertilizer Development Center (IFDC), Volume IV, Upper Volta.
[b]The range of per capita supply for 1970-79 is 148-181 kg. The average for the period is 167 kg/person. When adjusted for milling and other losses, the average supply is 150 kg/person.

TABLE 5.5
Estimated Food Requirements[a] in Burkina Faso

Crop	1970[b]		1980		1985		1990	
	Low	High	Low	High	Low	High	Low	High
Millet/ Sorghum	699	786	893	902	1,019	1,026	1,166	1,168
Maize	58	65	74	78	85	91	97	106
Rice	22	25	28	31	32	37	37	44
Wheat	23	26	29	33	33	41	38	50
Cowpea	109	123	139	145	159	168	182	196
Peanut	31	34	40	41	45	47	52	55

[a]In units of 1,000 metric tons.
[b]Estimates for 1970 are based on actual consumption. "Low" is based upon per capita consumption at estimated level of 1970 and "High" is based upon elasticity of demand (as per FAO projections). Taken from International Fertilizer Development Center (IFDC), Volume IV, Upper Volta.

TABLE 5.6
Cereal Imports[a] in Burkina Faso

Year	Wheat	Rice	Maize	Other	Grain Relief Aid	Total[b]
1960-65	8	3	1	2	0	14
1970-71	22	2	1	0	--	25
1972	34	2	6	1	--	41
1973	14	1	22	22	50	108
1974	21	3	24	30	95	170
1975	13	10	5	0	--	28
1976	16	12	1	0	--	29
1977	28	18	0	8	--	54
1978	24	10	0	29	--	63
1979	36	26	2	19	--	82
1980	50	29	3	13	--	95
1981	41[c]	20[c]	--	--	--	--

[a]In units of 1,000 metric tons per year.
[b]Excluding missing data.
[c]For only nine months.

variety, E-35-1, has a yield potential of 3.5 to 4.0 metric tons of grain per hectare, maize (IRAT[2] 100 and BDS III) 3.0 metric tons per hectare, and cowpea (KN1), 1.5 to 2.0 metric tons of grain per hectare on experimental plots. Even if only 50 to 60 percent of these yield levels are realizable under farm conditions, large shifts in production levels, and consequently in farming systems could result from the adoption of such new varieties.

Unfortunately, these potentials are not easy to realize. The grain producing farmers in Burkina Faso have not, as demonstrated by data in the following sections, adopted this technology. Commercial fertilizer use by small farms (10,000 tons per year) is insignificant. The production system in effect continues to follow traditional crop patterns and management practices. Questions regarding reasons for this have been raised by agronomists, economists, and policy makers. Is the current situation caused by technological relationships, economic feasibilities or lack of knowledge and resources needed to translate the various yield potentials into realities under real farm conditions and constraints?

Varietal improvements, more efficient agronomic practices, use of animal traction, and the use of modern farming practices are all needed. However, equally and perhaps most important is whether we can succeed in finding a suitable technology that is adaptable by current operators to the existing farm systems and which will increase production on a substantial number of the small farms that make up those systems.

In order to gain insight into this question it is important to consider the farming systems and methods used by small farmers in the three sample regions of Burkina Faso with major emphasis on crop production systems, factors influencing crop yields, use of modern inputs, animal traction and its impact on production and labor use, and some implications for research and development.

III. FARM PRODUCTION SYSTEMS IN THE STUDY REGIONS

The three rural development organization regions (ORDs), Ouagadougou, Ouahigouya and Zorgho selected for study in the first phase of the Farming Systems Research (1979 and 1980) are in the central region of the country. In terms of agricultural potential the Ouagadougou and Zorgho regions have been categorized as "poor" while the Ouahigouya region was categorized as a "very poor" region.[3] The three study areas have much higher population density (25 to 43 persons/sq km) than the rest of the country (average density 18 persons/sq. km). The pressure of population on agricultural land is accordingly highest in these regions.

The data presented in Table 5.7 provide a comparative view of the cropping systems and levels of productivity in the three study regions vis-a-vis some of the country's other regions in the selected regional development organizations, the ORDs.[4] These data indicate that cereal crops occupy the highest proportion of land under cultivation in all regions, although in the lower rainfall regions, the relative area under cereals is larger than in the high rainfall areas. For example, cereals occupy 92 to 93 percent of cultivated land in Dori located in the Sahel region (400-700 mm rainfall) and Yatenga (600-700 mm rainfall) regions of the Northwest Central Plateau as compared to 70 percent in the Western regions of Bobo, Diebougou and Banfora (1100-1400 mm). Similarly, in Ouagadougou, Yatenga

TABLE 5.7
Land Use, Cropping Systems and Yield Levels in the Three Study Regions vis-a-vis Some Selected Regions of Burkina Faso, 1977-1978

Area/Crop/Rainfall	Study Regions/ORDs			Other Selected ORDs				
	Ouaga	Yatenga	Koupela	Bobo	Diebougou	Banfora	Fada	Dori
Total Cultivated Land Area 1977 (1,000 ha[a])	490	220	130	150	200	90	190	140
Area Under Cereals (1,000 ha)	390	205	100	105	140	70	155	130
Percent of Area Under Cereals	80	93	77	70	70	78	82	93
Cotton (1,000 ha)	4	--	--	20	4	--	--	--
Legumes (1,000 ha)	15	9	18	11	16	8	17	--
Cultivated Area per Active Person (ha)	0.96	0.80	1.00	1.02	1.10	--	1.02	0.74
Area Per Farm Under:								
Sorghum (ha)	3.64	1.80	2.70	2.46	1.80	--	2.95	2.30
Millet (ha)	3.64	1.20	2.70	1.26	1.90	--	1.48	2.30
Maize (ha)	--	0.005	0.13	0.30	0.30	--	0.42	--

Peanut (ha)	0.15	0.15	0.56	0.42	0.45	--	0.77	--
Cowpea (ha)	--	0.20	0.15	0.24	0.65	--	--	0.77
Cotton (ha)	0.09	0.03	0.03	0.72	0.11	--	0.30	--
Yields Per Hectare (1977-78):								
Sorghum (kg[b])	495	368	650	844	545	560	848	148
Millet (kg)	408	300	360	690	434	520	618	229
Maize (kg)	263	206	250	1,045	651	850	1,230	--
Peanut (kg)	315	313	500	620	402	780	718	250
Cotton (kg)	365	201	229	866	249	140	700	200
Rainfall (mm):[c]								
Minimum	750	600	700	1,100	1,100	1,200	700	400
Maximum	1,000	700	1,000	1,200	1,200	1,400	1,000	700

[a]One hectare (ha) = 2.47 acres.
[b]One kilogram (kg) = 2.20 lbs.
[c]Rainfall in milimeters (mm).

Source: Ministry of Rural Development, Government of Burkina Faso.

and Koupela (Central regions) sorghum and millets are comparatively more important in cropping patterns than they are in the Western and the Eastern regions. In the Western and the Eastern regions, maize, peanut and cotton occupy a more important place than in the Central region.

Also noticeable are significant inter-regional yield differentials for the major crops such as millet and sorghum. Per hectare yield of millet is as low as 229 kg in Dori, 300 kg in Ouahigouya (Yatenga), 408 kg in Ouagadougou region as compared to 690 kg in Bobo and 618 kg in Fada. These differences are consistent with rainfall patterns. Other crops evidence similar yield differences. The average per hectare yield for cereal crops in the country is estimated to be 500 kg.

Differences in yield reflect, among other things, conditions of rainfall, soil fertility, management practices, and the overall resource endowments of the various regions. Equally important, they may suggest future possibilities and prospects for productivity-increasing efforts through technological changes, and developmental policies with regard to infrastructures, credit and fertilizer distribution, and farmer training and skill formation programs. This is especially true of differences in yields among farmers of the same and/or relatively homogeneous regions. There are cases in other countries where traditional farm management specialists have simply carried the "best" of the local practices from one farmer to another.[5] Researchers have isolated and developed varieties and methods to make the high yields easier to achieve. Incentives and infrastructure needs were isolated in the process and considerable economic development achieved at relatively low cost.

Cropping Systems

Millet and sorghum are the two most important cereal crops produced and consumed by farmers in Burkina Faso. As shown by the data in Table 5.8, millet, the principal field crop occupies 66.4 percent of the total cropped area followed by sorghum with about 20 percent. The other cereal crop grown almost universally by farmers is maize, although in terms of its relative share in the total cropped area it occupies only about 3 percent of the total farmed land. Peanuts are an important cash crop which is grown on 7.6 percent of the total cropped land in the study regions. In addition, there are a number of minor crops such as okra, bambarra nuts, roselle, etc. that are grown either as sole crops or as associated crops. In terms of land area, such crops occupy between 1 and 2 percent of total cropped land.

TABLE 5.8
Crop Area Distribution in Sample of Small Farms, Burkina Faso, 1980

Principal	Under Crops in Sample		Average Cropped Area
Crops[a]	Area (ha[b])	Percent	Per Household (ha)
Millet	203.58	66.4	3.40
Sorghum	60.89	19.8	1.00
Maize	8.63	2.8	0.14
Peanut	23.34	7.6	0.34
Bambarra Nuts	3.96	1.3	0.07
Okra	0.73	0.2	0.01
Misc. Crops	5.64	1.9	0.09
Total	306.77	100.0	5.05

[a]Of the field areas, 96 to 98 percent are under millet and sorghum and associated crops with cowpea as the most dominant second crop in association. Millet and Sorghum are also grown as associated crops.
[b]One hectare = 2.47 acres.

Source: Farming Systems Research, Sample Survey, 1980.

Growing crops in association is an important characteristic of the existing farming system that is practiced universally by small farmers in most parts of Africa. Sometimes farmers grow four to five crops in the same field. Data in Tables 5.9 and 5.10 show the crop associations followed by farmers in the study regions. Cowpea is by far the predominant second crop grown in association with cereal crops such as millet and sorghum. In fact, cowpea is grown mostly as an associated crop. Cowpea as a sole (mono) crop is more the exception than the rule on small farms in West Africa.

There are at least two hypotheses regarding the practice of growing crops in association in preference to mono cropping. Some crops are more susceptible to insects when grown in pure stands. Secondly, cowpea, the most important associated crop is a legume with some nitrogen fixation effects. However, this effect may be small with the low proportion of cowpeas in the usual crop mix.

TABLE 5.9
Cropping Patterns: Percentage Distributions of Fields by Crops, Sample of Small Farms in Three Regions, Burkina Faso, 1980

	Nedogo		Aorema		Digre	
	Main Fields	All Fields	Main Fields	All Fields	Main Fields	All Fields
Millet Mono	13.3	5.0	9.2	2.7	1.0	0.4
Millet Cereal	6.6	2.5	--	--	5.4	1.8
Millet Cowpea	63.9	24.0	69.3	20.0	91.4	29.5
Millet Others	16.2	6.0	21.5	6.2	2.2	0.7
	100.0	37.5	100.0	28.9	100.0	32.4
Red Sorghum Mono	27.3	2.5	--	--	4.3	1.0
Red Sorghum Cereal	15.2	1.7	--	--	2.9	0.7
Red Sorghum Cowpea	33.3	2.8	--	--	84.1	20.3
Red Sorghum Others	24.2	2.2	--	--	8.7	2.0
	100.0	9.2	--	--	100.0	24.0
White Sorghum Mono	10.0	0.6	--	--	50.0	0.7
White Sorghum Cereal	5.0	0.3	--	--	--	--
White Sorghum Cowpea	85.0	4.7	100.0	13.0	25.0	0.4
White Sorghum Others	--	--	--	--	25.0	0.4
	100.0	5.6	100.0	13.0	100.0	1.5

Maize Mono	9.3	1.0	22.2	1.8	7.7	0.4
Maize Cereal	37.2	4.4	50.0	4.0	84.6	4.6
Maize Cowpea	--	--	--	--	7.7	0.4
Maize Others	53.5	6.3	27.8	2.2	--	--
	100.0	11.7	100.0	8.0	100.0	5.4
Peanut Mono	46.0	8.0	48.3	12.4	50.0	9.5
Peanut Cereal	--	--	--	--	--	--
Peanut Cowpea	--	--	--	--	5.6	1.0
Peanut Others	54.0	9.4	51.7	13.3	44.4	8.4
	100.0	17.4	100.0	25.7	100.0	18.9
Okra	95.0	10.2	40.7	2.2	85.7	4.2
Okra Others	5.0	0.5	59.3	3.2	14.3	0.6
	100.0	10.7	100.0	5.4	100.0	4.8
Bambarra Nuts	28.0	2.0	57.1	10.7	45.5	3.5
Bambarra Nuts & others	72.0	5.0	42.9	8.0	54.5	4.2
	100.0	7.0	100.0	18.7	100.0	7.7
Roselle and Others	--	--	--	--	--	0.7
Rice (Paddy)	--	0.8	--	--	--	1.4
Cowpea	--	--	--	--	--	0.4
Cowpea and Others	--	--	--	--	--	--
Other Crops	--	0.3	--	--	--	2.8
Red Pepper	--	--	--	--	--	--
	--	100.2	--	99.7	--	100.0

Source: Farming Systems Research, Sample Survey, 1979-80.

TABLE 5.10
Relative Distribution of Fields by Crop Combination, Sample of Small Farms in Three Regions, Burkina Faso, 1980

Crop Combination	Percent Distribution of All Fields Operated by Households		
	Nedogo	Aorema	Digre
Millet Mono	5.0	2.7	0.4
Millet and Cowpea	3.0	8.4	2.2
Millet and Roselle	5.0	6.2	0.4
Millet and Red Sorghum	1.0	--	0.7
Millet and Earthpea	0.3	--	--
Millet, Cowpea, and Roselle	20.5	--	15.0
Millet, Bitto, and Cotton	0.6	--	--
Millet, Red Sorghum, and Cowpea	0.6	--	14.0
Millet, White Sorghum, and Roselle	0.6	--	0.4
Millet, Cowpea and Rice	0.3	--	--
Millet and Other	0.8	11.7	0.7
Red Sorghum Mono	2.5	--	1.0
Red Sorghum and Maize	0.8	--	--
Red Sorghum and White Sorghum	0.6	--	--
Red Sorghum, White Sorghum, and Roselle	0.3	--	0.7
Red Sorghum and Cowpea	1.0	--	4.0
Red Sorghum, Cowpea, and Roselle	2.5	--	14.0
Red Sorghum, Cowpea, and Sesame	0.3	--	--

Red Sorghum and Roselle	1.0	--	1.0
Red Sorghum and Others	--	--	1.0
White Sorghum Mono	0.6	--	3.3
White Sorghum and Cowpea	0.8	--	--
White Sorghum, Cowpea, Millet, and Roselle	0.3	4.0	--
Maize Mono	1.0	2.0	0.4
Maize and Red Sorghum	0.8	--	0.7
Maize and White Sorghum	1.3	--	--
Maize and Roselle	1.3	2.2	0.4
Maize and Other	7.0	4.0	4.0
Peanut Mono	8.0	12.4	10.0
Peanut and Roselle	9.0	2.7	5.2
Peanut and Other	0.3	10.7	4.8
Okra Mono	10.0	--	4.4
Okra and Other	0.6	5.4	0.7
Rice Mono	0.8	--	--
Bambarra Nuts Mono	2.0	10.4	3.6
Bambarra Nuts and Roselle	5.0	4.8	1.8
Bambarra Nuts and Other	--	3.0	2.5
Other Crops	0.3	--	2.5
Total	100.0	100.0	100.0

Source: Farming Systems Research, Sample Survey, 1979-80.

In most cases farmers in the sample followed a continuous crop rotation pattern, i.e., millet (and associated crops) after millet, and sorghum after sorghum with minor adjustment with changes in peanut planting. Farmers have followed this practice for decades without any application of fertilizers. Despite the cultivation of cowpea as a legume crop in association, soil fertility has definitely been depleted over time. Farmers have tried to avoid this problem to some extent in some areas of the Mossi Plateau by leaving land fallow. However, the practice of fallowing has been limited considerably by the increasing pressure of population on land.

Cowpea as stated earlier is grown universally as an associated crop with millet and sorghum. In the sample of 50 farmers selected for intensive observation, there was only one who grew cowpea as a single crop and that also in only one of his fields which constituted 0.4 percent of the number of fields operated by the farmer.

Since cowpea production as a single crop is not common, it could be difficult to promote the idea of a single cowpea crop at this stage. In most of the crop research and field trials it seems to have been generally assumed that farmers will grow cowpea, or other crops such as sorghum or millet, as a single crop not in association with other crops. Farmers in general do not follow such a practice, nor do they generally accept such a system.

Maize is grown by every household with an average of one to two fields per household. However, only a small proportion of cropped land is devoted to this crop. Maize is usually grown on land closest to the compound (*champs de case*). Fields close to the compound are generally of better quality in terms of soil fertility. Farmers have, over time, augmented the fertility of the soil in these fields with household and other forms of organic waste materials.

Peanuts are produced by almost all households because of the cash value to the household. This crop occupies 7 to 8 percent of total farm land and is of greater economic importance than the rest of the minor crops in the household's farm production system. As explained later, women play an important role in the production of this crop.

Labor Supply on Small Farms

Estimates of the available labor supply are presented by village in Table 5.11. On an average, a farm household in the sample has 4 to 6 labor units available for work. Next in importance to land, the amount of available labor determines

TABLE 5.11
Estimates of Available Labor Force,[a] Sample of Small Farm Households in Three Regions, Burkina Faso, 1980

	Nedogo		Digre		Tanghin		Aorema		Sodin	
	No.	%	No.	%	No.	%	No.	%	No.	%
Male Adults Per Household	1.8	33	1.7	41	1.5	42	1.8	41	3.2	52
Female Adults Per Household	2.8	51	2.0	48	1.6	44	2.2	51	2.2	35
Male Children Per Household	0.4	7	0.20	5	0.40	11	0.30	7	0.35	6
Female Children Per Household	0.5	9	0.25	6	0.10	3	0.05	1	0.45	7
Total Male and Female Children Per Household	0.9	16	0.45	11	0.50	14	0.35	8	0.80	13
All Labor Force Per Household[b]	5.5	100	4.2	100	3.6	100	4.4	100	6.2	100
Average Size of Household[b]	11.3		11.0		11.0		13.4		15.2	

[a]This estimate is based on the following conversion ratios:
1 man's labor = 1 labor unit;
1 female's labor = 0.75 labor unit;
1 child's labor (10-14 years of age) = 0.50 labor unit.

[b]Absentee members of households are excluded from the table.

Source: Farming Systems Research, Sample Survey, 1979-80.

household farm production because of the dominance of human labor in the current production systems. Agriculture in Africa is a highly labor intensive industry which caters to the subsistence needs of rural farm households. Under the existing production technology which is basically a land and labor using technology with minimal or no use of modern capital, the available labor supply plays a crucial role in determining the quantity of land that households can farm, and the timeliness of the various operations necessary to realize crop production. Labor shortage in a rather land abundant and capital scarce system of production can seriously constrain production.

A fairly substantial part of the household labor supply is comprised of women and children in the household. Women and children provide 50 to 66 percent of the total labor available to households. Amounts of labor used in farm activities will be considered in later sections.

Another characteristic of the crop production system on small farms is the role of women in managing crop fields.[6] For crops such as peanuts, bambarra nuts (*pois de terre*), okra and roselle (bitto), women, mainly the household head's wife or wives play an important role in managing production and sales of the household (Tables 5.12 and 5.13). In all but one of the sample villages, all the bambarra nut fields were farmed by women. Eighty-four to 100 percent of the fields and 43 percent of the land area in okra, 100 percent of roselle fields, and 48 to 76 percent of the fields and 51 percent of the land are in peanut were under the management of women in the households. The women also produced cereal crops. The number of millet fields farmed by women accounted for 23 to 36 percent of all millet fields and 6 percent of the land area under millet. Twelve to 40 percent of all sorghum fields and 4 percent of the area under sorghum were in charge of women.

Although women play an important role in production and marketing operations, they have been bypassed by development and extension agencies. With respect to health, education and information systems, women are the most neglected segment of rural society. The use of animal traction and purchased farm inputs were confined to male (husbands) fields. Women in the sample (i.e., wives of household heads and other married women in the household) are 100 percent illiterate. They have little or no access to rural institutions because they are male dominated. Traditionally the husbands who are heads of households have kept their women away from such contacts.

Deficiencies in nutritional intake, environmental conditions and lack of basic health and clinical services, are reflected in

TABLE 5.12
Distribution of Fields Under Selected Crops by Household Member Classification, Sample from Three Regions of Burkina Faso, 1980

Crops	Percent of Fields								
	Head of Household			Head's Wives and Other Female Members			Head's Sons and Other Male Members		
	Nedogo	Aorema	Digre	Nedogo	Aorema	Digre	Nedogo	Aorema	Digre
Millet	49	62	71	36	23	24	15	15	5
Red Sorghum	79	--	40	12	--	40	9	--	20
White Sorghum	75	60	81	20	30	11	5	10	8
Maize	98	89	80	--	--	7	2	11	13
Peanut	24	19	39	68	76	48	8	5	13
Okra	3	8	--	97	84	100	--	8	--
Bambarra Nuts	--	--	36	100	100	59	--	--	5
Roselle	--	--	--	100	--	100	--	--	--
Rice (paddy)	100	--	25	--	--	--	--	--	75
Cotton	--	--	--	--	--	--	--	--	--
Cowpea, Bambarra Nuts	--	--	100	--	--	--	--	--	--
Red Pepper	--	--	100	--	--	--	--	--	--
Other Crops	--	--	--	--	--	--	--	--	--

Source: Farming Systems Research, Sample Survey, 1979-80.

TABLE 5.13
Percentage Distribution of Cropped Area Under Major Crops in the Sample Regions, Burkina Faso, 1980

Crop	Percentage of Cropped Area Managed by: Male Members	Female Members
Millet	88.7	11.3
Sorghum	95.5	4.5
Peanut	48.8	51.2
Corn	99.6	0.4

Source: Farming Systems Research, Sample Survey, 1979-80.

high infant mortality. By the survey estimate, 33 percent of the children born per wife die while young (before they reach 5 years of age). This is indeed a very high death rate among young children. The women (and children) in the household bear a major share of the burden of economic hardships caused by poverty.

Crop Yields

Table 5.14 presents per hectare yields of major crops estimated on the basis of total production divided by total area in the crop. These data include the kg harvested per hectare of the main crop, the associated crops and all crops combined.

As shown by the data in Table 5.14, on the average, farmers in the sample harvested 415 kg of crops per hectare from their millet fields, of which 376 kg was millet and the rest other cereals such as sorghum (20 kg), cowpea (15 kg), and miscellaneous crops (5 kg).

Per hectare yield from sorghum fields was estimated to be 571 kg--about 36 percent higher than the millet field yield. Of the per hectare production on sorghum fields, sorghum production accounted for 460 kg, cereals other than sorghum, 73 kg, cowpea, 30 kg, and miscellaneous crops in association, 9 kg. Sorghum is generally grown on better quality soils than is millet. Hence, soil differences may be important in explaining yield differences between these crops.

The fields with maize as the main crop yielded 1162 kg per hectare, the highest per hectare yield of all crop fields. Of the total, maize accounted for 960.5 kg, other cereals (sorghum

TABLE 5.14
Yields of Major Crops, Small Farms Sample From Burkina Faso, 1980

Crops and Their Associations	Average Yield for All Fields
Millet in Association:	
Millet (main)	375.9
Cereals (association)	20.0
Cowpea (association)	14.8
Others (association)	4.7
Total Yields	415.4
Sorghum in Association:	
Sorghum (main)	459.6
Cereals (association)	72.6
Cowpea (association)	29.7
Others (association)	9.3
Total Yields	571.2
Maize in Association:	
Maize (main)	960.5
Cereals (association)	146.0
Cowpea (association)	1.0
Others (association)	54.9
Total Yields	1,162.4
Peanuts in Association:	
Peanuts (main)	470.5
Cereals (association)	10.2
Cowpea (association)	0.0
Others (association)	38.4
Total Yields	519.1
Bambarra Nuts in Association:	
Bambarra Nuts (main)	331.4
Cereals (association)	0.0
Cowpea (association)	0.0
Others (association)	0.0
Total Yields	331.4

Source: Farming Systems Research, Sample Survey, 1979-80.

mainly) 146 kg, cowpea 1 kg, and other crops 55 kg. Maize is always grown on fields around the house, and generally receives the most manure and care. The fertility of soils around the compound is higher than in other fields. It is, therefore, reasonable to expect higher per hectare yields of maize. However, in terms of total crop area and production, maize occupies a very small place in household farm production activities.

Peanut yields averaged 470.5 kg per hectare. In addition, peanut fields yielded about 49 kg of other crops grown in association.

Overall, the per hectare yields estimated for the major crops demonstrate very low productivity conditions on small farms. This is a major factor in the domestic supply of food crops.

IV. HOUSEHOLD GRAIN PRODUCTION, MARKETED SURPLUS AND PRICING

A household in Mossi villages generally consists of the head of the household, his wife, or wives, and their young children. Sometimes married sons and other relatives also live in the family compound of the chief of the household. The average size of a household in the sample is about 12 persons present in the household, and 15 including the absent (migrant) members. Land and labor are the two most important resources of small farm households in African agriculture. Land distribution has generally been governed by local tribal customs and traditions. Individual rights and ownership follow a well defined system but are quite different from Western systems, or those found in several other Anglophone countries in the third world.

The capital of small farm operators consists mostly of small hand tools and implements used for planting and weeding operations. Animal traction is not universal in Burkina Faso. The Farming Systems Research survey indicated that its use at present is limited. Farm production is heavily dependent on labor because most of production activities are carried out by hand.

Grain Production

The average size of farm households operated in the sample is 5.05 hectares (12.5 acres) of cropped area which is less than a half hectare (1 acre) of land area per person in the household. Given the existing low farm productivity (420 kg to 572 kg of grain per hectare), a farm size of 5 hectares is very small and may be inadequate to provide reasonable quantities of food for the family let alone any saving for further investment in agriculture. The estimated total production per sample household during 1980 consisted of 2.35 metric tons of crops (Table 5.15) of which millet accounted for 58.4 percent, sorghum, 23.4 percent, maize, 5.6 percent, cowpea, 3.4 percent, peanut, 6.8 percent, bambarra nuts, 1 percent, and the miscellaneous crops, 2.4 percent.

Based on the above estimates the per capita availability of food grains and dry pulses is 156 kg per annum, while for all crops it is about 172 kg. However this is the total available food grain supply of the households assuming no marketing. Households do sell some of the crops to meet their cash obligations. This is estimated to be between 10 percent to 15 percent of total production.[7] Thus, if one were to take out the quantities sold by households, the available food grain supply per person per annum will be reduced to 120 kg of food grains and dry pulses, and 146 kg of all crops produced by households.

TABLE 5.15
Household Crop Production: Small Farms Sample, Burkina Faso, 1980

Crop	Quantities Produced Per Household (kg)	Quantity Available Per Person (kg)[a]
Millet	1,370.0	100.0
Sorghum	548.4	40.0
Cowpea	80.4	6.0
Maize	134.0	10.0
Peanut	160.0	11.6
Bambarra Nuts	23.3	1.9
Okra	8.0	0.5
Miscellaneous	22.0	1.7
All Crops	2,346.1	171.7

[a]Per capita available to household equals total production minus 15% to account for grain loss and seeds, etc. divided by the number of persons in the household.

Source: Farming Systems Research, Sample Survey, 1979-80.

Marketed Surplus

Although most of the food grain produced by small farm households is for self consumption, farmers do engage in marketing activities. Almost every household, whether operating a small farm or a big farm (note the average farm size in the sample is a little over 5 hectares, about 12.5 acres, of cropped land area), sells some quantities of food grain, no matter how small. According to our estimates for the sample households, farmers marketed in the range of 9 to 11 percent of their total grain output (for the country as a whole, the World estimates 15 percent as the marketed surplus). For the 1980 harvests, the average prices received by farmers for the major commodities that they sold to provide traders in open (outside of ORDs) markets were as follows: millet, 65 CFA francs per kg; sorghum, 60 CFA francs; maize, 65 CFA francs; peanut, 85 CFA francs; and cowpea, 95 CFA francs. These estimates are based on the weekly data on prices gathered from selected rural markets for the period June 1980 to January 1981. Note that during the same period, the announced official prices in Burkina Faso were 40 to 45 CFA francs per kg for millet, sorghum and maize--the three most important cereal grains in the country! In fact, there had been a little change over the last three years in the announced official prices at which the Rural Development Organizations, ORDs, were required to buy from the farmers (32 CFA/kg in 1977-78, 40 CFA/kg 1978-79, and 40-45 CFA/kg 1980-81).

The foodgrain market in Burkina Faso is only a small fraction of total production. Most of the farmers not only sell, although small, but also buy grains from the market particularly during the lean months (during the planting and weeding seasons). Prices are generally higher during the months when farmers buy food from the market than in the post-harvest months when they sell their produce in the market. The farmers' limited holding capacity (to retain foodgrains) puts them to a disadvantage at both ends: they sell when prices are low and buy when they are high. The other, and quite related, phenomenon in foodgrain marketing is that the government in Burkina Faso, like most other governments in Africa, has intervened, particularly since 1973, in an attempt to secure low-priced food for urban consumers. In 1973, a National Grain Agency (OFNACER) was created under the Ministry of Commerce with the objective of regulating foodgrain prices and building buffer stocks. The Regional Development offices (ORDs), created under the Ministry of Rural Development, has had the responsibility of purchasing foodgrains for the state agency (OFNACER) from the farmers at the announced official prices. However, it has been difficult for the government to implement its monopolistic foodgrain purchasing policy; prices have continued to fluctuate, and farmers have continued to sell their produce to private traders and merchants. According to the official estimates for 1978-79, the relative shares of OFNACER, village cooperatives and private traders in the total purchase of cereals from the farmers were as shown in Table 5.16.

The other way through which the government has attempted to manipulate domestic prices is cheap imports of cereals. The overall result of the two-fold government intervention policies has been lower prices of grains received by farmers. Domestic food prices have generally been lower than the world prices. The case of cash (export) crops is equally, perhaps more, lamentable in regard to government's pricing policy.[8] In most cases, such policies have tended to hurt rather than help agriculture, and hence the need for serious thinking in the direction of what Professor T.W. Schultz (1964, 1979) calls "incentive prices."

V. INPUT-USE, CREDIT CONSTRAINT AND DEMAND FOR PRODUCTION CREDIT

Use of Modern Inputs

Cereal crops are grown under traditional farming practices which in general do not include the use of modern inputs such as high yielding seed varieties, chemical fertilizers, and modern means to control insects, pests and diseases. The use of such

TABLE 5.16
Relative Shares of Cereal Purchases from Farmers

Agency	% Share	Per kg Value (Price)
OFNACER	46.2	40 CFA francs[a]
Village Cooperatives	26.8	56 CFA francs
Private Traders	27.0	44 CFA francs

[a]200-225 CFA francs = 1 US$.

modern inputs on the sample farms appears minimal (Table 5.17). Low levels of application were used and only on parts of the fields. A difference in yield was usually observed between the fertilized and unfertilized portions. In all, 7.5 percent of the sample farms applied some phosphate with a total expenditure of about $12 per farm that used this fertilizer. In the case of sorghum, 2.5 percent of the sample farmers used phosphate with an average expenditure of $16 per farm. It is important to note that in all cases, farmers received this fertilizer at government subsidized prices from Purdue's Farming Systems Research Unit conducting trials in farmers' fields. Otherwise, farmers would not have used any fertilizers for these crops.

Per farm expenditure on chemical insecticides and fungicides was $0.50 to $1.50 based on the four farmers in the entire sample of 50 that used such chemicals. Fifteen percent of the millet producing farmers and 5 percent of the sorghum farmers reported to having used some chemicals. For all the purchased inputs, the per hectare expenditure is estimated at $0.50 for millet and sorghum and about $6 for maize.

Except for a few farmers who tried new crop varieties (mainly sorghum and cowpea) under the supervision of experiment station scientists, the farmers in the sample grew local

TABLE 5.17
Fertilizers and Pesticides Used By Sample Farmers,[a]
Burkina Faso, 1980

Fertilizer and Pesticide Category	Percentage of Farms		Average Value (US $) Per Farm Using Inputs	
	Millet	Sorghum	Millet	Sorghum
Phosphate	7.5	2.5	12.00	16.00
Cotton Fertilizer	2.5	--	5.00	--
Organic Manure	30.0	10.0	Not Estimated (home produced)	
Pesticides/Insecticides/Fungicides	15.0	5.0	1.50	0.50

[a]These estimates are based on selected farmers in the three sample villages (Nedogo, Digre, and Tanghin).

Source: Farming Systems Research, Sample Survey, 1979-80.

varieties. Some of the local varieties are fairly drought-resistant. It may be possible to achieve higher yields from some of these varieties with the use of fertilizer and moisture conserving management practices.

Most production activities on the majority of farms are carried out manually with small farm tools and equipment that have been in use for several hundred years. In the entire sample, only 33 percent of the households have animal traction, and 90 percent of these use a donkey to pull the hoe or cultivator. Of the sample villages, Nedogo in the Ouagadougou region which is close to the capital city has the largest fraction (60 percent) of sample households with animal traction. In the other sample villages, this ranges from 10 percent to 40 percent of total farm households.

Credit Constraint and The Demand for Production Credit

The credit survey was confined to 40 farmers in three villages. The main purpose of this survey was to find out: (a) if the farmers felt the constraint of credit on agricultural production (the question actually asked was, "During the last year, was there anything that you wanted to do to improve farm production but you could not for lack of money"?); (b) if the farmers wanted to borrow credit, if available, for the purpose of agricultural production during the year in question (1980); and (c) if yes, for what purposes (uses) and from which sources? These questions were repeated in the other sample villages to confirm and verify the results of the three-village survey. The summary of the main findings is presented in Table 5.18. As evident from the results, 82 percent of the farmers felt that credit was a constraint on agricultural production, and the same percentage of them would have liked to borrow credit, if it was available. The major source from which farmers (88% in the sample) *wanted* to borrow was the semi-government organizations called ORDs, primarily because of low interest rates (8 to 10%) and a sense of security. The commercial banks, or even the cooperatives, were not chosen as the primary source for borrowing credit by farmers. The cooperatives, friends and relatives, the Farming Systems Unit, and other (not specified) each accounted for 3 percent of all sources from which farmers wanted to obtain credit.

It is indeed instructive to note that the major constraint, as felt by the farmers, to efforts directed towards increasing crop production was the lack of improved farm implements, such as cultivators, seeders, etc. and draft animals--oxen and donkeys

TABLE 5.18
Credit Constraints, Credit Demand, Uses and Sources: Sample Farmers' Responses, Burkina Faso, 1980

Credit Constraint and Demand	Percentage
Farmers experiencing credit constraint on production	82.0
Farmers demanding credit for production purposes	82.0
Uses for Which Credit Demanded	**Percentage**
Purchase of Farm Equipment (cultivators, seeders, etc.)	54.8
Purchase of Oxen	16.1
Purchase of Fertilizers, Seeds, etc.	12.9
Purchase of Donkey	6.5
Purchase of Cart	6.5
Trade and Commerce	3.2
	100.0
Preferred Sources of Borrowing	**Percentage**
Rural Development Organizations (ORDs)	88.0
Cooperatives	3.0
Commercial Banks	0.0
Relatives/Friends	3.0
Farming Systems Research Unit (of Purdue)	3.0
Others	3.0
	100.0

Source: Farming Systems Research, Sample Survey, 1979-80.

to pull the plows. About 55 percent of the farmers wanted to use the credit borrowed for the purchase of farm equipments, while 22.6 percent, for the purchase of draft animals (oxen and donkeys, Table 5.18). The next important use for which credit was sought was the purchase of fertilizers and seeds. The data in Table 5.18 clearly suggest that the farmers in Burkina Faso have serious credit constraints on their production efforts, that they feel the need for borrowing credit, and that they are prepared to borrow and use it on productive purposes. At present the farmers in general are outside of the organized credit market; and the lack of credit, like the lack of agricultural extension services, is bound to come in the way of agricultural improvement in the country. The problem is not that the farmers do not have the genuine demand for credit, or for farm equipments, or for other inputs. The real problem appears to be that credit is unavailable to these farmers and there is a lack of a well supervised production-credit arrangement or system that could help promote technological change in traditional agriculture. The present condition, in fact, severely limits the spread of technological innovations, however limited these innovations are.

VI. PRODUCTION FUNCTION FOR MAJOR CROPS: RESULTS OF REGRESSION

Production relationships using log linear functions were estimated for the four major crops: millet, sorghum, maize and peanut. The dependent variable in the equations was per hectare yield of the principal crops with the *parcelle* (field) as the unit of observation. This allowed for more degrees of freedom, although household characteristics other than land, labor and input use could not be incorporated into the model. However, the estimated relationships enable us to measure the contribution of production factors such as field size, labor, associated crops, input use, and animal traction on farm yields. The variables of the production model are specified in Tables 5.19 through 5.23.

Impact of Field Size

The impact of the land variable, the size of the parcelle (field) on per hectare yield is consistently negative for all the four crops and in all the estimating equations of the production model. The regression coefficient for this variable is negative and statistically highly significant in all cases (Tables 5.19 through 5.23). This means that as the size of the parcelle or field increases the per hectare yield tends to decline. The relationship is plausible under the present farming system. In most cases the largest fields operated by households are located farthest from the compounds and village fields and these fields are inferior in terms of soil fertility. Such fields are given the lowest priority in regard to timely performance of operations

TABLE 5.19
Estimated Regression Coefficients for Millet:
Log-Linear Function

Independent Variables of the Production Model	Equation (1)[a,b]	Equation (2)[b,c]	Equation (3)[a,d]
Size of Parcelle (hectare)	-.4905***[e] (.0520)[f]	-.4978*** (.0521)	-.4507*** (.0507)
Relative Yield of Cereals (Association)	-.0940** (.0449)	-.0918** (.0449)	--
Relative Yield of Cowpea (Association)	-.0840** (.0332)	-.0834** (.0331)	--
Relative Yield of Other Crops (Association)	-.0844*** (.0260)	-.0848 (.0260)	--
Land Preparation Labor (hours)	.0159 (.0311)	.0165 (.0334)	.0290 (.0311)
Planting Labor (hours)	.1085 (.0552)	.1080 (.0556)	.1071 (.0551)
1st Weeding Labor (hours)	.0707 (.0583)	.0778 (.0588)	.0663 (.0584)
2nd Weeding Labor (hours)	.0354 (.0264)	.0385 (.0279)	.0902** (.0261)
Use of Animal Traction Dummy, 0-1	.3771** (.1320)	.3794*** (.1316)	.2672*** (.1289)
Input Expenses CFA francs	.0042 (.0264)	.0032 (.0263)	-.0181 (.0265)
Constant "a"	4.1843	--	4.4746
R^2	.2607	.2643	.2258
$\bar{R}^2$	.2370	.2407	.2086
F	11.00	11.20	13.13
N	323	323	323

[a]All labor hours weighted equally regardless of source, i.e. male, female, or children.
[b]Dependent variable = logarithm of yield (kg) per hectare of millet.
[c]Labor hours weighted: 1 male labor hour = 1 labor hour; 1 female labor hour = .75 labor hour; 1 child labor hour = .50 labor hour.
[d]Dependent variable = logarithm of yield (kg) per hectare of all crops (millet and associated crops).
[e]Statistical significance levels: *** = 1%; ** = 5%; * = 10%.
[f]Figures in parentheses are standard errors.

such as planting and weeding. The households first try to plant and/or weed the better quality fields located closer to the compounds and the village and then move to the big fields that are located farther away. It may also be more difficult to manage labor and other inputs on the more distant large fields.

The size-yield relationship as estimated in this study on the basis of per hectare yields and size of field or parcelle does not imply anything about farm size as related to efficiency. The question being addressed is how farm production per hectare is influenced through the use of yield augmenting inputs and other factors. In some cases animal traction may lead to somewhat larger farms. However, in regions of high population density, e.g., the Mossi plateau, high quality land may not be available for more extensive farming.

Impact of the Yields of Associated Crops

Increased associated crop yields tended to have negative effects on the yields of the major crops with the exception of cowpeas in association with maize. This is indicated by preponderance of significant negative coefficients for associated crop yields (Tables 5.19, 5.20, 5.21, 5.22).

These negative coefficients for the associated crop yields imply a competitive relationship between the main crop and the associated crop in question. In the case of cowpeas with maize, the significant positive coefficients indicate a complementary relationship.

To test further for complementarity between the crops for a range on the production possibility curve, quadratic production functions for sorghum and millet were estimated. The results (Table 5.23) do not show any strong complementarity among the crops grown in association with these major crops.

TABLE 5.20
Estimated Regression Coefficients for Sorghum:
Log-Linear Function

Independent Variables of the Production Model	Equation (1)[a,b]	Equation (2)[b,c]	Equation (3)[a,d]
Size of Parcelle (hectare)	-.4791***[e] (.0809)[f]	-.4921*** (.0811)	-.4966*** (.0821)
Relative Yield of Cereal Crop (Association)	-.3558*** (.0812)	-.3536** (.0808)	--
Relative Yield of Cowpea (Association)	-.3244 (.0631)	-.3179 (.2200)	--
Relative Yield of Other Crops (Association)	-.0724 (.0665)	-.0748 (.0663)	--
Land Preparation Labor (hours)	.0472 (.0556)	.0582 (.0596)	.0266 (.0564)
Planting Labor (hours)	.0571 (.1036)	.0703 (.1039)	.1051 (.1055)
1st Weeding Labor (hours)	.0092 (.0932)	.0010 (.0942)	.1163 (.0958)
2nd Weeding Labor (hours)	.0507 (.0458)	.0526 (.0483)	.1041** (.0461)
Animal Traction Dummy, 0-1	.0585 (.2204)	.0590 (.2200)	-.1886 (.2265)
Input Expenses CFA francs	-.0769 (.0551)	-.0792 (.0551)	-.0922 (.0567)
Constant "a"	4.3338	4.2854	4.3863
R^2	.4471	.4212	.2645
$\bar{R}^2$	.4064	.4109	.2275
F	11.00	11.18	7.14
N	147	147	147

[a-f]Footnotes as on Table 5.19, except sorghum for millet.

TABLE 5.21
Estimated Regression Coefficients for Maize:
Log-Linear Function

Independent Variables of the Production Model	Equation (1)[a,b]	(2)[b,c]	(3)[a,b]
Size of Parcelle (hectare)	-.5862**[d] (.1090)[e]	-.6012** (.1088)	-.7832** (.1556)
Relative Yield of Cereal Crop (Association)	-.0750 (.1483)	-.0636 (.1486)	--
Relative Yield of Cowpea (Association)	1.6656*** (.2771)	1.6360*** (.2756)	--
Relative Yield of Other Crops (Association)	-.1389 (.1037)	-.1544 (.1030)	--
Land Preparation Labor (hours)	-.1135 (.1505)	-.0940 (.1490)	.0545 (.1786)
Planting Labor (hours)	-.3002 (.1695)	-.3387* (.1682)	-.2860 (.2046)
1st Weeding Labor (hours)	.4687** (.1569)	.4729* (.1609)	.4610* (.1869)
2nd Weeding Labor (hours)	.2229** (.0859)	.2402* (.0908)	.2169* (.1038)
Animal Traction Dummy, 0-1	.4028 (.2724)	.4191 (.2701)	.2327 (.3269)
Input Expenses CFA francs	-.0250 (.0803)	-.0342 (.0796)	-.0231 (.0972)
Constant "a"	4.4504	4.4188	3.7075
R^2	.6338	.6366	.4019
$\bar{R}^2$	.5774	.5807	.3404
F	11.25	11.38	6.53
N	76	76	76

[a-c]Same as for Table 5.19, except maize for millet.
[d-e]Same as e-f for Table 5.19.

TABLE 5.22
Estimated Regression Coefficients for Peanut:
Log-Linear Function

Independent Variables of the Production Model	Equation (1)[a,b]	(2)[b,c]	(3)[a,d]
Size of Parcelle (hectare)	-.4211***[e] (.0623)[f]	-.4404*** (.0679)	-.4890*** (.0715)
Relative Yield of Other Crops (Association)	-1.0389 (.7033)	-1.0572* (.6902)	--
Land Preparation Labor (hours)	.1574** (.0585)	.1921** (.0605)	.1338** (.0613)
Planting Labor (hours)	-.0127 (.0849)	-.0326 (.0839)	.0303 (.0891)
1st Weeding Labor (hours)	.1736** (.0659)	.1954** (.0665)	.1785** (.0674)
2nd Weeding Labor (hours)	-.0898 (.1213)	-.1059 (.1359)	-.0417 (.1273)
Animal Traction Dummy, 0-1	.1425 (.1456)	.1695 (.1431)	.1351 (.1528)
Input Expenses CFA francs	.1070 (.2071)	.1125 (.2123)	.1338 (.2173)
Constant "a"	4.0049	3.0376	3.8718
R^2	.2119	.2342	.2198
$\bar{R}^2$	.1784	.2016	.1909
F	6.3190	7.1881	7.6060
N	197	197	197

[a-f]Footnotes as on Table 5.19, except peanut for millet.

The situation investigated involves only existing systems of production for both sole cropping and associated cropping of the parcelles. The relative yield situation might change with varieties and production technology.

Yield-Labor Relationships

For millet, the relationship between yield and labor use for land preparation, planting and weeding operations is positive (Table 5.19). Although the coefficient of labor input is relatively small, it is statistically significant at the 5 percent level for planting labor, and at the 10 percent level for weeding labor, but not significant for land preparation labor. This indicates that the use of more labor would increase yields for this major crop. It is true, however, that the marginal productivity of labor is very low. The marginal productivity of labor in sorghum production is positive but low (Table 5.20). Also, it is not significant statistically.

For maize, both the land preparation and planting labor has negative but non-significant coefficients (Table 5.21). However, yield is strongly positive and significantly related with weeding labor time. The coefficient of maize weeding labor is much larger in magnitude and higher in the level of statistical significance than for all of the other crops.

For peanut, land preparation and weeding labor has a positive influence on yields. The estimated coefficient of the labor input in both cases is statistically significant at the 5 percent level (Table 5.22).

Animal Traction

This variable was treated as a "dummy" with 0-1 values (zero for no animal traction). Hence, the coefficient of this variable indicates change in the level of yields for any given input combination, i.e., change in the Y-intercept of the logarithmic form of the model without changing the other coefficients of the production function. In all the production models the coefficient of the animal traction dummy appears positive (Y intercept--constant 'a' ± the positive dummy coefficient). This means that other things constant, animal traction farms will harvest higher levels of yield than the farmers without animal traction. It also appears that the use of animal traction enables households to farm more land. The principal constraint to area expansion seemed to be timely first weeding, and animal traction apparently overcomes this constraint and enables land area expansion.

TABLE 5.23
Estimated Regression Coefficients For Millet and Sorghum:[a] Quadratic Function

Independent Variables of the Production Model	Millet Coefficient	Sorghum Coefficient
Size of Parcelle (hectare)	320.3249***[b]	-458.6395***
	(16.7698)[c]	(43.0311)
Relative Yield of Cereal (Association)	-110.3155	-43.1831
	(67.2628)	(27.5320)
Cereal Crops, Relative Yield Squared	36.3340	1.6631
	(25.7473)	(2.1016)
Cowpea, Relative Yield (Association)	-96.9536	4.4499
	(57.6148)	(88.4837)
Cowpea, Relative Yield Squared	12.0186	-20.7892
	(9.2607)	(24.5289)
Other Crops, Relative Yield (Association)	-102.0852	41.8141
	(209.5350)	(136.3184)
Other Crops, Relative Yield Squared	76.8300	-5.4259
	(147.7320)	(25.3404)
Land Preparation Labor (hours)	.5680**	-.5298
	(.2722)	(.7693)

Planting Labor (hours)	.9298**	-.3816
	(.2642)	(.5126)
1st Weeding Labor (hours)	-.0598	-.1994
	(.0826)	(.1834)
2nd Weeding Labor (hours)	.0052	.2528
	(.0692)	(.2072)
Animal Traction Dummy, 0-1	45.0259	89.5023
	(30.6806)	(54.3913)
Purchased Inputs CFA francs	.0069	-.1222*
	(.0276)	(.0678)
Constant "a"	-3.5711	43.4027
R^2	.8197	.6286
$\bar{R}^2$	.8123	.5923
F	108.072	17.32
N	323	147

[a]Dependent variable = logarithm of total production (kg) of millet or sorghum per parcelle.
[b]Statistical significance levels: *** = 1%; ** = 5%; * = 10%.
[c]Figures in parentheses are standard errors.

A more complete treatment of the animal traction technology variable, as well as a much greater in-depth study of the effects of this variable based upon rigorous econometric analyses are provided in Chapter 6.

Overall Performance of the Production Models

On the basis of the multiple coefficient of determination, R^2, the production function models leave a large part of the variation in yields unexplained. In the case of millet, none of the three estimating equations explains more than 23 to 26 percent of total variation in yield. For sorghum and maize, the performance is a little better with R^2 in the range of .41 to .63. For peanuts, the value of the R^2 ranges between .18 and .20. In all cases, however, R^2 values are statistically highly significant.

On the other hand, with production in semi-arid Africa so heavily dependent upon weather conditions, it may not be possible to formulate a weather variable. Also, the production functions used in the present analysis were not intended for predictive purposes. The above analysis was made to give evidence that crop yields in the sampled region were responding to certain important controllable input variables and to show the direction and the relative size of these responses.

VII. THE ESTIMATED MARGINAL VALUE PRODUCTS (MVP) OF PRODUCTION INPUTS

The marginal value products (MVP) estimated from production functions for the four major crops are presented in Table 5.24.

The Marginal Value Product of Land

Maize land yielded the highest marginal value product of 25480 CFA francs (US $115) per hectare of area, followed by peanut land with a marginal value product of 22000 CFA francs (US $98) per hectare. Millet land gave the lowest marginal value product of 13422 CFA francs (US $60). Looking at the MVP figures, it would appear that farmers in the study region would be better off transferring land from millet and sorghum crops to maize and peanut production. However, this depends principally on four factors: (1) the availability of land suitable to maize production under the present conditions, (2) farmers' tastes and preferences, (3) the input supplies including the information systems required for growing maize and/or peanut,

TABLE 5.24
Estimated Marginal Value Products (MVP) of Field Size and Labor for Major Crops: Small Farms, Burkina Faso, 1980

Input	Millet	Sorghum	Maize	Peanut
Field Size (Hectares)	13422.0	14400.0	25480.0	22000.0
Planting Labor (Hours)	27.2	9.0	--	6.0
First Weeding Labor (Hours)	5.3	--	8.3	8.2
Second Weeding Labor (Hours)	4.0	8.3	--	--
Purchased Inputs (CFA francs)	0.85	--	--	--

[a]All values in CFA francs, 1 US$ = 225 CFA francs.

and (4) marketing and pricing. Furthermore, the availability of labor input can be a serious constraint in view of the relatively much higher priority currently assigned by farmers to millet and sorghum--the two most dominant crops in the existing farming systems which compete with maize and peanuts for labor and other inputs.

The Marginal Value Product of Labor

The marginal value product of planting labor is generally higher than that of weeding labor. However, overall the MVP of labor under the existing production system is extremely low. Except for planting labor of millet for which the MVP of labor per hour is 27 CFA francs (US $0.12), the MVP of labor estimated for the major production operations does not exceed 9 CFA francs per hour. Such low marginal value products indicate the low value of additional labor in the current farming system. Wages of hired labor are usually no higher than its marginal value product otherwise no labor is hired. The fact that little, if any, labor is hired on a wage rate basis in present day (traditional) agriculture as practiced in the region is consistent with this finding.

The Marginal Value Product of Purchased Inputs in Millet Production

The marginal value product of purchased inputs with current production practices was estimated for millet. However, there was an insufficient number of users to give a reliable estimate, and hence, no information for this crop has been provided in the table.

VIII. HOUSEHOLD LEVEL FARM PRODUCTION FUNCTION: THE MALE-FEMALE PRODUCTIVITY DIFFERENTIAL AND THE ECONOMIC EFFECTS OF EDUCATION[9]

In this section, the results of aggregate (household) level production ("earnings" or revenue) function analyses will be discussed to focus on two major issues: (i) whether there exists any significant difference between the male labor productivity and the female labor productivity on the farm; and (ii) whether education of farm people, although extremely low in general, has any effect on agricultural production, and what is the rate of return to schooling for farm households that operate small farms in a traditional farming setting.

The production functions formulated in this analysis are different from those formulated in earlier sections of this chapter. First, the estimated production functions in this section use aggregate (farm) level data on yields, revenue, and inputs such as labor, capital and land, instead of the individual crop data by fields as used previously and reported in Tables 5.19 through 5.24. Second, the dependent variable is measured in terms of revenue (income) earned by households by raising crops, instead of physical yields by crops and by fields. Third, the household variables such as education, farm capital assets, and the size of the farm, in addition to the labor input variables, have been incorporated into the estimating model whereas in the previous estimates of production function these variables could not be included as these characteristics are not associated with fields or parcelles of crops for which functional analyses were made. The variables of the estimated model are thus measured at the farm or household level. This, of course, has reduced the size of the sample. However, the estimates obtained through the aggregate (household) farm production function provide important insights into a few economic aspects of the production relationships that highlight the relative roles of the male and female labor inputs in production and differential in marginal productivity of the male-female labor in farming, and the economics of schooling in a rather static, traditional farming system.

The Estimating Model and the Results

Broadly, two forms of the income or "earnings" function were formulated, using the Mincer (1974) and Chiswick (1974) framework as a basis and making the usual assumptions. First a simple "earnings" (income) function with schooling as the only independent variable was specified as follows:

$$\ln E_{SCH} = \ln E_0 + rSCH + e \qquad (1)$$

where E_{SCH} = the earning (revenue or income) with schooling, SCH; r = the average rate of return on SCH; SCH = the measure of schooling (school years) completed; and e = the error term.

The annual "earnings" or income, E_{SCH} equals the sum of all individual crop yields multiplied by their respective prices for the gross revenue; while for the net revenue, it equals the gross revenue minus the expenses on purchased inputs. Estimates were made using, alternatively, the gross revenue (earnings) and the net revenue (earnings) as the dependent variable. The schooling variable, SCH, has been measured in two ways: (i) the schooling of the household head (the number of school years completed), and (ii) the (total) schooling of all members of the household (the total school years). Note that in the sample all the female members (wives of household heads and other females) were illiterate, thus the total schooling variable represents the schooling of only the male members of households.

It is important to state that in the above formulation of the "earnings" or income function, the coefficient of schooling, r, measures the average rate of return to schooling. In the above formulation (1), it is also evident that the effects of variables (inputs) other than schooling on household income have been left out. This is not realistic, however. The other variables like land and labor are not constant across households; also, these may well be correlated with schooling, SCH. Hence, the desirability of expanding the simple model (1) to incorporate those major inputs, besides schooling, that generate household (farm) income. Considering the structure of the sample households, an expanded income model was, therefore, specified as follows:

$$\ln E_{SCH} = b_0 + b_1 \ln FRM + b_2 \ln MAL + b_3 \ln FEM + b_4 KAP + b_5 SCH + b_6 DAT + e \quad (2)$$

where ln FRM = log of the amount of land farmed (farm size in hectares); ln MAL = log of male labor hours worked (over the production period); ln FEM = log of female labor hours worked (during the production period); ln KAP = log of value of farm capital (in CFA francs); DAT = dummy for animal traction taking the value of 1 if animal traction was used by household and 0 otherwise.

In the above formulation the coefficient of schooling (SCH) can be treated as an approximation to the rate of return to schooling. However, as argued by Welch (1970), the estimate of this coefficient (b_5) is closer to what he calls the "worker" or "direct" effect of schooling. The coefficient b_5, however, reflects the contribution of schooling to income when all other inputs are held constant whereas an important ("allocative") component of the return to schooling is in an efficient allocation of resources. This (allocative) aspect of the role of schooling is better reflected by the coefficient (r) of schooling estimated through Equation 1.

The Results of Estimates

The results of the estimated coefficients on the variables are presented in Table 5.25. Ordinary least-squares (OLS) method was used to estimate the coefficients; and, hence, the usual assumptions and limitations will hold in case of the present analysis of the regression analysis. First, the focus of discussion is on the result pertaining to the effect of labor inputs on income (farm production), and the male-female labor productivity differential.

The Male-Female Productivity Differential

Given the specification of the earnings function in model (2), it is evident that the coefficients of the variables representing hours of work by males and females are also elasticities of income (output) with respect to the labor input in these two categories. The results of estimates of the coefficients using alternative specifications of the model are presented in Table 5.25 (Equations 1, 2, 5 and 6). As shown by the estimated coefficients on male and female labor inputs, there is an enormous difference in the elasticities with respect to male and female labor. Although a comparison of the two elasticities itself is instructive, the estimated marginal products would be somewhat more relevant statistics. Using the elasticities in Equation 5 (Table 5.25), marginal products for the male and female hours at the sample mean are, respectively, about 112 and 683 (units of the local currency CFA francs). The difference between the two marginal products, although not as large as between the elasticities, is still large. It shows that, at the sample mean, an hour of female work is nearly *six* times as productive as an hour of male labor. Note that the difference in the two marginal products is *not* just a reflection of the levels of usage of the two inputs. In fact, as the difference between the elasticities also suggests, the average female hours worked are *higher* than the average male hours.

The results of a large and significant differential between the male and female labor productivity on the farm remained consistent and stable across several alternate specifications of the model estimated. Although quite a few researchers have given favorable accounts of female productivity in African countries in several economic activities (for example, farming, trade and commerce), such a large productivity differential has not been mentioned, perhaps not even suspected. For example, Boserup (1970, pp. 16-22), a renowned researcher on women, observed "Africa is a region of female farming par excellence." She also states that farm female workers in general work longer hours and even today "village production in Africa South of Sahara continues to be predominantly female farming." That women work longer hours and work harder than the male family members in the labor-intensive farm production systems of countries South of Sahara has been recorded also by other researchers cited by Boserup, and by Standing and Sheehan (1978), Durand (1975), and Singh (1981). However, almost all observations pertaining to women's productivity in farming are based on survey data on the hours worked and not on any measures of productivity.

The results of estimates of productivity differentials as reported in Table 5.25 are quite consistent with the descriptive accounts of most researchers but go far beyond such remarks about the long hours worked or "hard labor" of females in the region. It is not easy to give sharp explanations for the observed difference which is huge. However, some possible

TABLE 5.25
The Estimated Coefficients of Earnings (Income) Model Using Ordinary Least-Squares (OLS), Burkina Faso, 1980

Variables[a]	Equations for:							
	Total Earnings (Revenue)				Net Earnings (Revenue)			
	(1)	(2)	(3)	(4)	(5)	(6)	(7)	(8)
Log of Land-Area Farmed (ln FRM)	.6166 (6.89)[b]	.6384 (6.39)	--	--	.6047 (6.23)	.6190 (5.84)	--	--
Log of Male Labor Hours (ln MAL)	.0288 (.41)	.0701 (.92)	--	--	.0231 (.31)	.0643 (.79)	--	--
Log of Female Labor Hours (ln FEM)	.2384 (3.49)	.2050 (2.62)	--	--	.2545 (3.43)	.2221 (2.64)	--	--
Log of Value of Farm Capital (ln KAP)	.0415 (1.31)	.0458 (1.31)	--	--	.0441 (1.29)	.0486 (1.30)	--	--
School Years Completed by the Head (SCH)	--	.0315 (1.41)	.0979 (2.82)	--	--	.0304 (1.27)	.0980 (2.76)	--
Total School Years Completed by all Members of Household (SCH)	.0685 (3.48)	--	--	.0746 (2.08)	.0694 (3.25)	--	--	.0754 (2.05)

Animal Traction Dummy (DAT)	.0909 (.89)	.0271 (.23)	--	--	.0970 (.88)	.0342 (.27)	--	--
Constant "a"	9.55	9.60	11.77	11.69	9.50	9.51	11.76	11.68
R^2	.80	.75	.12	.07	.77	.73	.12	.07
F	28.97	22.43	7.93	4.31	24.98	19.68	7.60	4.21
N (no. of households)	51	51	59	59	51	51	59	59

[a]The dependent variable E_{SCH}, earnings of households, has been defined in terms of: (a) total earnings = the sum of all the individual crops raised on the farm x their respective prices and (b) net earnings = total earnings as in (a) minus the expenses on purchased inputs. Earnings are in CFA francs; 1 US$ = 225-250 CFA francs (1980-81).

[b]t-statistics in parentheses.

explanations could be offered. A part of the explanation may lie in the difference in age-composition of the two sexes, particularly the household heads and their wives who contribute a major share of farm labor. For example, the average age of the household heads in the sample was 57 years while that of the wives in the sample was 40 years. The wives are typically much younger than men among the household residents. The proportion of the working husbands (household heads) above 50 to the total household heads was .67, while the corresponding proportion of heads' wives was .23. Another possibility is that the practice of polygyny may encourage greater "competition" among wives and might thus make them more productive than males. Yet another possible explanation may lie in that male hours are, for some reason, systematically overreported relative to the females. For example, the hours reported as worked on the farm by males might really be hours spent on other chores, and might thus not be reflected in farm income. Nevertheless, the explanations suggested appear conjectural although the observed male-female productivity differential is probably solidly based in the data. In view of the limitations of the cross-sectional nature of the analysis, caution needs to be exercised while drawing any strong conclusions. However, further exploration of the male-female productivity difference in the region seems to offer a promising area for additional research.

Education, Farm Production and Economic Returns to Schooling

The question of the economic role of education in traditional farming systems has long been discussed and debated. Much of the received wisdom on the economics of education suggests a low payoff to schooling in a traditional production setting, and further suggests the "allocative" effect of schooling to be particularly small in such a static environment (Welch 1970; Schultz 1975). The present analysis of the household level data on production and schooling will help in making judgements on these aspects, and thus may add to the evidence on the subject.

The results of estimates obtained through both model (1) and (2) presented in Table 5.25 are instructive for a discussion on the economics of schooling in farm production and on the returns to schooling. First and a significant result of the estimated coefficient on schooling is the strongly positive effect that this variable exercises on farm production and income even in a dominantly traditional and rather static production setting. The coefficient on schooling, whether of all the members of the household, or of the household head, remain stable and consistent across several alternative specifications of the estimating equation. Second, the estimated coefficient on total schooling in the household (Equations 4 and 8, Table 5.25) suggest a rate of return of the order of 7 to 7.5 percent, which appears plausible. Third, and quite interestingly, the schooling of the household head, the chief decision maker in the household, yields a higher rate of return (of 10 percent) than that on the other members' schooling. In all equations (3, 4, 7 and 8 of Table 5.25) the coefficient on schooling is statistically significant at least at the 5 percent level.

Fourth, the results highlight the allocative and direct effects of education in this setting. The coefficient on schooling (SCH) estimated through Equations 1, 2, 5 and 6) can still be treated as an approximation to the rate of return on schooling. However, as argued by Welch (1970), the estimate of the coefficient is closer to the "worker" or "direct" effect of schooling; the coefficient reflects the contribution of schooling to income when other inputs are held constant. Interestingly, the rate of return on the schooling of all household members, due to the "direct" effect, which is about 6.9 percent (Equation 1), is just a little lower than the total effect of 7.5 percent reflected in Equation 4. This implies that the allocative effect of education for most members is small and this may be expected from the static character of the environment and the limited allocative role of members other than the household head.

It is instructive to examine, in this context, if the results of the estimated coefficient on schooling of the head of the household differs from that of the other members of the household. This involves comparison of the results reported in Equations 2, 3, 6, and 7 which are for the head, and Equations 1, 4, 5 and 8 which are for other members of the household (Table 5.25). Note that the coefficient on schooling estimated through Equations 2 and 6 for the household head indicating the direct or "worker" effect of schooling is much smaller than the coefficient on schooling of other members of the household estimated through Equations 1 and 5. This is what may be expected, because the primary role of the household head is decision making (allocative) and his direct contribution to farm work (as worker) is relatively limited. There is a comparatively larger difference between the estimated coefficient on schooling (SCH) in (2) and (3) and/or in (6) and (7) for the household head than, for example, between the estimated coefficient on schooling in (1) and (4) and/or in (5) and (8) for other members of the household. These results suggest, as one would expect, a relatively larger "allocative" return on the schooling of the household head than that on the other members.

The major conclusions following from the results of the estimates of the production (earnings) functions with schooling may be summed up as follows: First there is a large productivity differential between the male and female labor on the farm, an hour of female labor appears 5 to 6 times as productive in farming as an hour of male labor. A number of possible explanations have been attempted for such a large differential; however, there is a great need for further research in this rather challenging area. Second, the overall rates of return to education appear broadly of the order of 7 to 10 percent. Considering the low (below primary) levels of schooling in the sample, and considering, at the same time, the very traditional character of the production setting, the rate of return appears plausible. Third, the allocative return to the schooling of the household head seems quite large (larger than the direct return), while the "worker" effect of the head's education seems small. Such a pattern is reasonable to expect since the major role of the household head in farming is decision-making or "allocative." Fourth, for the household members other than the head of the household, the primary gain from schooling seems to be accruing through the "direct" or "worker" effect in Welch's terminology. This is so because of the relatively static production system and the limited decision-making (allocative) role for the other family members than the household head.

Overall, in whatever way one may look at the results, the effect of education, the human capital variable, emerges significant on agricultural production and household income.

NOTES

1. ICRISAT = International Crop Research Institute for Semi-Arid Tropics, Regional Office, Burkina Faso.

2. IRAT = Institute for Research in Tropical Agriculture

3. Singh, Ram D., Major Cropping Patterns in SAFGRAD Countries and Government of Upper Volta, Ministry of Planning and Rural Development Annual Reports.

4. There are in all 11 ORDs (Regional Development Organizations) which are geographic units covering the country.

These are autonomous organizations responsible for extension services, credit, marketing and rural infrastructures.

5. This system was used extensively by the extension services in the U.S.A. Farm records systems and farm tours have this comparative aspect as one of their functions.

6. For more details regarding the role of women see Margaret O. Saunders, "The Mossi Farming System of Upper Volta." FSU Working Paper No. 3, OUA/CSTR--Joint Project 31 between USAID and Purdue University, April 1980.

7. Based upon the report obtained from the World Bank and the Government of Burkina Faso.

8. See for an interesting discussion of governmental intervention in marketing and pricing of agricultural commodities, Robert Bates, "States and Political Intervention in Markets: A Case Study From Afria," Agricultural Economic Workshop Paper No. 82; 5, Feb. 11, 1982, Department of Economics, University of Chicago.

9. This section draws material from a jointly authored paper (with Rati Ram) titled: "Farm Households in Rural Burkina Faso: Some Evidence on Allocative and Direct Return to Schooling and Male-Female Productivity Differentials," forthcoming in *World Development* 6(1988). Some portions are taken directly from this paper.

REFERENCES

Bates, Robert H. "States and Political Intervention in Markets: A Case Study from Africa." Agricultural Economics Workshop Paper #82:5, The University of Chicago, Department of Economics, Chicago, 1982.

Boserup, F. *Women's Role in Economic Development*. London: George Allen and Unwin, 1970.

Chiswick, B.R. *Income Inequality*. New York: Columbia University Press, for the NBER, 1974.

Christenson, P. *Farming System Unit (FSU) Field Trials in Sample Villages, 1979-80, SAFGRAD/FSU Report*. West Lafayette: Purdue University, 1980.

Durand, John D. *The Labor Force in Economic Development*. Princeton: Princeton University Press, 1975.

Eischer, C.K. and D.C. Baker. "Research on Agricultural Development in Sub-Saharan Africa: A Critical Survey." Michigan State University Paper No. 1., Department of Agricultural Economics, East Lansing, Mich., 1982.

Food and Agricultural Organization (FAO). *Yearbook*. Rome: FAO, 1975-1980.

Government of Upper Volta, Annual Agricultural statistics, 1978-79.

Government of Upper Volta, Ministry of Rural Development, Annual Reports.

International Agricultural Development Services. *Agricultural Development Indicators--A Statistical Handbook*. New York: International Agricultural Development Services 1978.

International Crop Research Institute for Semi-Arid Tropics (ICRISAT). *Annual Reports*. Ouagadougou, Burkina Faso: ICRISAT, 1980-81.

International Fertilizer Development Center (IFDC). "West-Africa Fertilizer Study, Upper Volta." Vol. IV, Technical Bulletin IFDC T-6, March 1977.

International Institute for Tropical Agriculture (IITA). *Annual Reports*. Lagos, Nigeria, 1979, 1980.

Institute for Research in Tropical Agriculture (IRAT). *Annual Reports*. 1979, 1980, 1981.

Lang, Mahlon, Ronald Cantrell and John Sanders. "Identifying Farm Level Constraints and Evaluating New Technology in the Purdue Farming Systems Project in Upper Volta." Paper presented at Farming Systems Symposium, Kansas State University, Manhattan, Kansas, October 31, 1983.

Mincer, Jacob. *Schooling, Experience and Earnings*. New York: Columbia University Press, 1975.

Psacharopoulus, George. "Returns to Education: An Updated International Comparison." In *Education and Income*. World Bank Staff Working Paper No. 402, The World Bank, 1980.

__________. *Returns to Education: An International Comparison*. New York: American Elsevier Jassey-Bass, 1973.

SAFGRAD/FSU. *1982 Annual Report*. IE&R and IPIA, Purdue University, 1983.

Saunders, Margaret O. "The Mossi Farming System of Upper Volta." FSU Working Paper No. 3. West Lafayette: Purdue University, 1980.

Schultz, T. W. "The Value of the Ability to Deal with Disequilibria." *Journal of Economic Literature* 13(1975):827-846.

__________. "The Economics of Being Poor." Nobel Lecture, The Nobel Foundation, Stockholm, Sweden, December 10, 1979.

__________. *Transforming Traditional Agriculture*. New Haven & London: Yale University Press, 1964.

Singh, Ram D. "Major Cropping Patterns in SAFGRAD Countries, Upper Volta." Document #7, SAFGRAD/FSU, Ouagadougou, 1981.

________. "Small Farm Production Systems in West Africa and Their Relevance to Research and Development." Agricultural Economics Workshop Paper No. 81: 20, The University of chicago, Department of Economics, May 28, 1981.

Standing, G. and Glen Sheehan, eds. *Labor Force Participation in Low Income Countries*. Geneva: International Labour Office, 1978.

Welch, Finis. "Education in Production." *Journal of Political Economy* 78(1975):35-59.

World Bank. *The World Development Reports*. Washington, D.C.: The World Bank, 1981, 1982, 1983 and 1984.

________. *Accelerated Development in Sub-Saharan Africa: An Agenda for Action*. Washington, D.C.: The World Bank, 1981.

________. "Upper Volta Agricultural Issues Study." Report No. 3296-UV, Washington, D.C.: The World Bank, 1982.

6

The Economics of Animal Traction

I. INTRODUCTION

In predominantly traditional farming systems that prevail in most parts of Africa, as emphasized earlier, human labor plays a crucial role in agricultural production. Among the several bottlenecks, labor bottlenecks are, to quote the World Bank (1981), "a key constraint to agricultural progress in Africa," and, furthermore, "a breakthrough in ox-drawn cultivation," says the World Bank, "would obviously have the most powerful effect on labor productivity." This does not imply that the labor constraint is the only or the most important constraint on agricultural production in the region. However, it needs to be stressed that development policy in the African region should focus on measures that increase labor productivity, in particular the use of farm implements in addition to other improved farm practices such as the introduction of drought-resistant high-yielding varieties of seeds and the use of fertilizers. Although the use of animal traction has been recognized and often stressed as an important technological innovation in the setting of a highly labor intensive production system (Barrett 1982; Singh 1981), not much progress has been achieved in this respect (World Bank 1981). similarly, the semi-arid, rain-fed, cereals, such as sorghum and millet have remained by and large unaffected by the so-called Green Revolution. Despite the inflow of foreign assistance, including the establishment of several regional agricultural research centers, very little seems to have gone to farmer's field with any major impact on yields of cereal crops in West African countries (Singh et al. 1984).

In recent years, one of the major objectives of a development program of the national planning agencies of several African nations is to increase agricultural production and rural incomes through a number of schemes, the introduction of animal traction (AT) being an important one of these. The use of animal draft power is assumed to enhance farm productivity by alleviating major labor constraints. The overall likely results are: (a) an increase in labor productivity, (b) relieving of seasonal labor bottlenecks, and (c) increase in crop yields. However, despite the importance of animal traction program in the region, the few available studies have lent rather inconclusive evidence resulting into a debate about whether animal traction and its introduction in the setting of a highly labor-intensive subsistence farming system of Africa is, or is not, a technological innovation with a positive effect on farm productivity and income.

Spencer and Byerlee (1976) maintain that the technical change necessary to raise labor productivity in agriculture would entail the adoption of some form of labor-saving technology. This is in line with the argument pertaining to technological change in LDCs agriculture (Hayami and Ruttan 1971). Spencer and Byerlee (1976), in a study of rice production in Sierra Leone, support the hypothesis that mechanical technology increases labor productivity, while the introduction of bio-chemical technology increases the productivity of land. In more recent studies, Singh et al. (1984) and Barrett et al. (1982) agree, although partially, on the positive association between the use of AT and total agricultural output. However, Barrett et al. question the relationship due to the effect of farm size in terms of the number of workers. Similarly, the findings of Delgado and McIntire (1982) show no increase in acreage with animal traction except when differences in family size are ignored. According to Jaeger (1984) weeding with animal traction tends to increase labor productivity, and this could allow significant acreage increases where land is not a constraint. Overall, however, these studies seem to cast doubt on the significance of the effects of animal technology (AT) on the productivity of land, in particular with regard to the food grain crops such as millet and sorghum. The question concerning whether AT is an innovation technology, or whether it is merely a simple substitute for labor, was also investigated in these studies, although the answer appears to be ambiguous. Due to ambiguous and even conflicting conclusions and/or inferences that could be drawn from previous empirical studies, doubts may be raised about the economic potential of animal traction in the West African region.

The continuing controversy over the economic benefits and at the same time the growing emphasis on promoting the use of animal traction in farming indeed point to the need for adequate documentation and additional analytical work on the subject. Animal traction is being used profitably by farmers in other parts of the third world countries, particularly Asia, and there appears to be a growing demand for it in Africa (World Bank 1981; Singh 1984). The need for empirical research is much greater now than it was ever before because of the increasing focus on public supported projects related to animal traction in the Sahel. In the World Bank's view "more emphasis should now be placed on measures that increase labor productivity, in particular, the use of farm implements, ox-drawn cultivation." It is, therefore, of interest to further evaluate the economic effects of this (animal traction technology) on agriculture in the region. Using farm level data the study reported in this chapter examines: (i) the association between the use of animal traction and per acre yield; (ii) the relationship between animal traction and the farm size; and (iii) extent to which this is a labor saving innovating technology, or whether it is a simple substitute for labor.

The effects of an animal traction program on the productivity of farms are naturally correlated with some other factors such as the size of the farm, the size of the household, the cropping patterns, and the use of purchased inputs--improved seeds, fertilizers and/or pesticides. In order to reduce the degree of correlations of these factors, the study will use the average concept variables such as the total yield per hectare, the input expenditure per hectare, or the total yield per worker or per labor hours. Since the markets for crop products and input factors are not well established in this particular rural area, the study will use the labor and animal traction inputs and the output variables measured by physical units in order to prevent distortion of the results from the analysis.

The rest of the chapter is devoted to the discussion of the following: In part II, a brief description of the crop production system, the use of animal traction on the small farms and the demand for animal traction is provided, while in part III, the major results of the findings on the economic effects of animal traction, are presented. The summary and conclusions are contained in part IV.

II. THE USE OF AND DEMAND FOR ANIMAL TRACTION IN CROP PRODUCTION

This study focuses on four major cereal crops produced in Burkina Faso: millet, sorghum, corn (maize) and peanuts. Millet and sorghum contribute more than 85 percent of the total cultivated area, while peanuts contribute 7.6 percent, and corn 2.8 percent, respectively.

Most production activities on the majority of farms are carried out manually with small farm tools and equipment that have been in use for several hundred years. In the entire sample, only 33 percent of the households had animal traction, and 90 percent of these used a donkey to pull the hoe or cultivator. Of the sample villages, Nedogo in the Ouagadougou region which is close to the capital city had the largest proportion (60 percent) of sample households using animal traction. In the other sample villages, this ranged from 10 percent to 40 percent of total farm households. Where animal traction was used for the major crops, the cropped area on which animal traction was used varied from about 31 percent (peanut and maize) to 40 percent (millet) of the total cropped area under the respective crops (Table 6.1).

The data in Table 6.2 also show that not all of the farmers owning animal traction have used it uniformly for comparable farming operations. For example, animal traction was not used by the sample farmers for either land preparation or for planting activities of major crops such as millet and sorghum in Nedogo, the village with the largest percentage of sample farmers with animal traction. In the other villages the percentage of fields for which animal traction was used by the sample farmers for land preparation and planting was rather small. However, for crops such as maize and peanuts, farmers used animal traction for land preparation in 38 to 39 percent of fields in Nedogo, and 17 to 50 percent of fields in Aorema (Table 6.2). Animal traction was used for weeding in most of the villages under study, but not on all fields. Note that it was used in weeding for almost all major millet and sorghum fields.

The currently increasing demand for animal traction by the West African farmers is partly an attempt to increase farm size. Although farms are small in terms of land and the use of modern inputs by Western standards, the relatively large sized farms have greater need and also more resources for the use of animal traction than the small sized farms. The first advantage of animal traction that a farmer points out[1] is that it helps him

TABLE 6.1
Use of Animal Traction on Small Farms, Sample Households, Burkina Faso, 1980

Principal Crops[a]	Total Cropped Area in Hectares[b]	Percent of Cropped Area on Which Animal Traction was Used
Millet	203.6	40.3
Sorghum	60.9	33.4
Maize	8.6	30.9
Peanut	23.3	31.4
Bambarra Nuts	4.0	20.7
Okra	0.7	6.8
Misc. Crops	5.6	16.3

[a]96 to 98 percent to the total cropped area under millet and sorghum had associated crops (secondary crops also grown in the field). Cowpea was the crop most used in such associations. Millet and sorghum were grown as associated crops as well as principal or main crops.

[b]Total area in the sample under various crops. One hectare equals 2.47 acres.

Source: Farming Systems Research, Sample Survey, 1979-80.

farm a larger land area. Besides, owning draft animals such as donkeys, oxen, and horses, and equipment, is a symbol of social prestige. Households owning such capital enjoy higher socio-economic status in the community. Hence, there is incentive to have such items even when not used to the fullest extent possible.

The growing demand for animal traction among the farmers in the study region has been demonstrated by the results of an attitude (opinion) survey conducted in four of the five sample villages. In response to the question, "During the last year, was there any measure that you wanted to undertake to improve agricultural production, but you could not for the lack of money (finances)?" 65 to 75 percent of the farmers interviewed answered "yes," and the major item on their agenda was "the purchase of draft animals and implements, in particular the ox-drawn plow." The majority of farmers now realize the usefulness of animal traction, but the constraint is that they cannot obtain it for the lack of financial resources. Fifteen to

TABLE 6.2
Use of Animal Traction For Farming Operations on Major Crops, Three Regions, Burkina Faso, 1980

	Percent With Animal Traction					
	(Ouagadougou)		(Ouahigouya)		(Zorgho)	
Operations	Farmers	Fields	Farmers	Fields	Farmers	Fields
Millet						
Land Prep.	--	--	70	25	--	--
Planting	--	--	10	2	--	--
Weeding	63	24	20	5	10	10
Sorghum						
Land Prep.	--	--	40	27	10	3
Planting	--	--	--	--	--	--
Weeding	53	46	--	--	10	3
Maize						
Land Prep.	32	38	44	39	--	--
Planting	--	--	--	--	--	--
Weeding	--	--	--	--	--	--
Peanut						
Land Prep.	33	17	60	50	--	--
Planting	6	2	--	--	--	--
Weeding	16	5	--	--	--	--
Farmers Owning Animal Traction (%)	60	--	40	--	10	--

SOURCE: Farming Systems Research, Sample Survey, 1979-80.

twenty years ago, the situation was much different. In one of the sample villages (in the Ouahigouya region), the village chief showed the survey team a storage facility located in the village where the government (ORD) had officials stored "hundreds of farm tools and implements, but the farmers never bought them." According to the village chief, "sometimes the tools were kept outside of the godown, and even the thieves would not take them!" The village chief's reasoning was that the farmers then did not see the benefits of using the improved farm implements. But now in the seventies and the eighties the condition seems to have changed; and the survey data show that it indeed has.

When questions were asked about the credit constraint and credit use, the majority of farmers surveyed said that the non-availability of credit most severely limited the purchase of trained and reliable draft animals, and "houe manga," a local name for a donkey drawn hoe first used in Manga, a village in Burkina Faso. More than 80 percent of the farmers in the credit survey wanted credit from formal credit institutions, the ORDs, the banks and cooperatives for the purchase of animal traction (both animals and draft equipment). Credit is unavailable to these farmers and trained animals similarly are relatively unavailable in the market. The absence of a well-integrated credit program that combines credit supply with supervision and production related information may be severely limiting the spread of technological innovations.

III. ECONOMIC EFFECTS OF ANIMAL TRACTION: MAJOR EMPIRICAL FINDINGS

The major economic effects of animal traction have been evaluated through: (i) descriptive analyses of the data on farm size, crop yields and labor use patterns vis-a-vis the use of animal traction on the sample farms, and (ii) regression analyses that provide estimates of the production (function) relationships at (a) the household or farm level, and (b) the field or parcelle level of crop production.

Descriptive Analyses: Farm Size Effects

The total amount of land area cultivated by households that employ animal traction is almost twice as large as for the (non-animal traction farm) households that employ hand cultivation methods (Table 6.3). However, this seems to be due primarily to the larger family size for households using animal traction (column 2, Table 6.3). Hence, this indicator may not be a precise indicator of whether a unit increase of labor input contributes a relative increase of farmed area in households that use animal traction compared to those households that employ hand cultivation methods. A better indication of the effect of animal traction on the farm size may be captured by measuring the land area cultivated per active worker (column 6). Measures of this kind show that households using animal

TABLE 6.3
Average Farm Size, Land Area and Use of Animal Traction, Burkina Faso, 1980[a]

	Number of Cases		Number of Household Members		Number of Active Workers[b]		Total Area Cultivated (hectare)		Land Area Per Household Member[c]		Land Area Per Active Worker[d]	
	NAT[e]	AT[e]	NAT	AT	NAT	AT	NAT	AT	NAT	AT	NAT	AT
Crop	(1)		(2)		(3)		(4)		(5)		(6)	
Millet	38	21	9.3	13.4	4.3	6.0	2.61	5.17	.32	.36	.67	.81
Sorghum	44	9	9.3	15.7	4.3	7.2	.94	2.50	.11	.17	.24	.39
Peanuts	49	6	10.4	11.8	4.8	5.2	.38	.72	.04	.07	.08	.15
Corn	48	5	10.4	11.5	4.9	4.8	.12	.26	.01	.02	.04	.06
Total	38	21	9.9	13.5	4.6	6.1	.92	3.41	.11	.24	.23	.54

[a]Figures are unweighted arithmetic means (averages)

[b]Number of active workers are counted as number of household members aged between 15 and 60 years old with the equivalent weight regardless of sex and age differences.

[c]Total area divided by the number of household members.

[d]Total area divided by the number of active workers.

[e]NAT stands for the sample of households using hand cultivation methods; AT for those using animal traction.

Source: Farming Systems Research, Sample Survey, 1979-80.

traction cultivated, on average, a significantly greater land area ranging from 87.5 percent for corn to 20.9 percent for millet. The data in columns 5 and 6 (Table 6.3) would support the contention that there is a positive connection between the size of the farm (i.e., land area farmed by the household) and animal traction. It is interesting to note that the positive effect of animal traction on size also holds for peanut and corn crops, which occupy relatively smaller fractions of the cultivated area. Hence, this result indicates that the use of animal traction may be a substitute for labor, independent of the scale effect.

Table 6.4 provides another indicator of the farm size effect of animal traction with respect to labor inputs. Overall, one hour of labor input for the households using animal traction cultivated about 30 percent more land area than a corresponding labor input for the (non AT) households employing hand cultivation methods. The difference in the average land area cultivated per unit of labor input is greatest in the case of peanut production and smallest in the case of millet production.

TABLE 6.4
Average Cultivated Land Area Per Unit Labor Hour (ha/hr), Burkina Faso, 1980

	Number of Cases		Ratio 1[a]		Ratio 2[b]		% Change Over NAT	
Crop	NAT[d]	AT[d]	NAT	AT	NAT	AT	(1)	(2)
Millet	38	21	2.24	2.35	2.56	2.69	4.91	5.08
Sorghum	44	9	2.45	2.86	2.86	3.33	16.73	16.43
Peanuts	49	6	1.58	2.58	1.96	3.11	63.29**[c]	62.83**[c]
Corn	48	4	1.47	1.99	1.63	2.37	35.37	45.50
Total	38	21	1.91	2.47	2.20	2.87	29.32	30.45

[a]Ratio 1 = 1000 x Land area/(Male Labor Hour + Female Labor Hour + 50% of Children Labor Hour).
[b]Ratio 2 = 1000 x Land area/(Male Labor Hour + 75% of Female Labor Hour + 50% of Children Labor Hour).
[c]Significant at the 5% level.
[d]NAT for non-animal traction farms; AT for animal traction.

Crop Yield Effects

As evident from the results presented in Table 6.5, the effects of animal traction on per hectare yields are quite mixed. For millet and sorghum crops, the effect turned out to be negative, while for peanut and corn crops, the effect was positive. The overall effect was negative. However, note that Table 6.5 shows relatively different sizes of average total land area for the four major crops. The size of land areas for millet and sorghum are greater than those for peanuts and corn for both subsamples of households. In the peanut and the corn cases, the animal traction program seems to have a positive effect on yield per hectare. On the other hand, in the millet and the sorghum cases, the animal traction program seems to have a negative effect on yield per hectare. The results may imply that, for cases of peanut and corn production, there exists an economies of scale effect due to the introduction of the animal traction. In contrast, for millet and sorghum production, it is likely that the negative effect of the law of diminishing returns is great enough to offset the effect due to economies of scale.

The question concerning the effect of animal traction on land productivity (the average per hectare yield effect) may be viewed from another perspective. This is that once the animal traction program is employed in a larger scale farm production (such as in millet or sorghum production) and it is fully

TABLE 6.5
Average Total Yield, Land Area and Yield Per Hectare, Burkina Faso, 1980

	Average Total Yield (kg)		Average Total Land Area		Yield (kg/ha)		
Crop	NAT[a]	AT[a]	NAT	AT	NAT	AT	% Change
Millet	1,234.3	2,300.1	2.61	5.17	473	445	-5.9
Sorghum	661.7	1,635.5	.94	2.50	704	654	-7.1
Peanuts	225.6	546.9	.38	.72	594	760	27.9
Corn	220.1	529.4	.12	.26	1,835	2,036	11.0

[a]NAT for non-animal traction farms; AT for animal traction.

utilized in the setting of the subsistence level farms, the effect of economies of scale may offset the effect of the law of diminishing returns. Then, an animal traction program could lead to an improved production technology and an increase in the productivity of land as well as labor.

Labor-Use Effects

Labor input per hectare of land area farmed and labor input per worker for a given unit of land area are higher for households employing traditional hand cultivation methods than those using animal traction for a broad range of labor operations, i.e., for tilling, weeding, and seeding. The results of this study presented in Table 6.6 strongly suggest that animal traction is a labor-saving technology. The data on land-labor ratios shows that labor unit per unit of land for the tilling operation for millet and sorghum production is higher for households that use animal traction (40.3 and 43.3 hr/ha, respectively) than for those households that employ hand cultivation methods (27.6 and 32.1 hr/ha, respectively).

If the level of technology is the same for the households using animal traction and for those households not using animal traction, then the inference that one may draw is that the animal traction program saves labor hours. However, the question as to whether the contribution of animal traction to

TABLE 6.6
Labor-Land Ratio and Labor Input Per Worker Per Unit Land Area, Small Farms, Burkina Faso, 1980

	Labor/Land Ratio (hr/ha)			Labor/Worker/Land Ratio[a]		
Labor	NAT[b]	AT[b]	% Change	NAT	AT	% Change
Tilling	177.4	61.1	-65.56	479.1	145.5	-69.64
Weeding	616.2	398.5	-35.33	6,265.7	884.2	-85.89
Seeding	125.5	81.5	-35.06	1,043.0	164.1	-84.27
Male	322.3	183.7	-43.00	2,361.9	483.2	-79.54
Female	487.1	289.5	-40.57	4,501.0	605.9	-86.54
Children	219.3	135.9	-38.03	1,849.8	209.5	-88.68

[a]In hours per worker per hectare.
[b]NAT for non-animal traction farms; AT for animal traction.

farm production takes the form of an innovative technology is unclear. The results of estimating the labor ratios for the four major crops are presented in Table 6.7. In the case of peanut and corn production, the output-labor ratios per unit land for the households using animal traction are about 2.5 times larger than for those using hand cultivation methods. However, in the case of millet and sorghum production, the difference between the output-labor ratios for the two groups of farms is relatively small, even though the major portion of animal power is allocated to these two crops.

Crop Production Relationships: Results of Regression

The small farm production function with animal traction variable were estimated using the ordinary least-squares (OLS) method. The form of the function specified for the purpose of evaluating the effects of the animal traction technology variable was of the Cobb-Douglas production function type. This assumed log-linear relationships between crop production and the explanatory variables included in the estimating model. In some cases, the regression coefficients were estimated using the weighted ordinary least squares.[2] The focus of the discussion will be on the results of the regression analyses of the crop production data and not on the methods or limitations of the statistical estimational procedures. The variables of the

TABLE 6.7
Labor-Land Ratio (hr/ha) for Four Major Crops,
Small Farms, Burkina Faso, 1980

	Millet		Sorghum		Peanuts		Corn	
Labor	NAT[a]	AT[a]	NAT	AT	NAT	AT	NAT	AT
Tilling	27.6	40.3	32.1	43.3	199.3	92.4	406.9	163.5
Weeding	542.5	427.8	631.5	418.2	513.3	282.5	765.6	374.4
Seeding	86.1	83.0	120.3	73.4	125.9	89.4	161.0	79.6
Male	242.5	212.5	289.2	206.9	185.5	55.5	555.5	172.0
Female	341.6	277.4	398.5	253.5	547.7	353.9	621.7	337.5
Children	144.2	122.5	192.6	149.2	210.2	109.6	312.8	215.8
Total	656.2	551.5	783.9	534.9	838.3	464.3	1333.4	615.7

[a]NAT for non-animal traction farms; AT for animal traction.

estimating production functions are described in each of the individual tables containing the results of estimates. Similarly, whether the estimated function relationships pertain to the individual crop(s) using the fields or parcelles (in which the individual crops are raised by households) as the unit of observation, or whether these are based upon the aggregate household (farm) level crop data (i.e., using the households or farms as the unit of observation), is also stated in the tables. Each table is, therefore, designed to be self-explanatory in terms of the specification and the number of variables included in the estimating model.

The Household (Farm) Level Crop Production Function

The results of the estimated production function using the aggregated household (farm) level data on the variables are

TABLE 6.8
Estimated Coefficients of Variables of the Household Level Production Function Using Ordinary Least Squares (OLS): Log-Linear Function

	Millet	Sorghum	Corn	Peanut
Labor (all operations)	.274*** (2.79)	-.016 (-.12)	.589 (4.67)	.253 (1.59)
Capital Assets	.014 (.64)	.299*** (2.79)	.098** (2.25)	.148* (2.34)
Animal Traction Hours	.028 (.56)	.086 (.90)	.141* (1.72)	.216** (2.34)
Intercept	4.254*** (6.83)	6.334*** (7.61)	1.021** (2.30)	4.237*** (4.31)
F	2.876**	2.730*	12.03***	4.284***
R^2	.155	.167	.474	.234
N (No. of Households)	51	45	44	46

[a]The dependent variable is the logarithm of yield (kg/ha).
[b](t-statistics in parenthesis)
[c]Statistical significance levels: *** = 1%, ** = 5%, * = 10%.

TABLE 6.9
Estimated Coefficients of the Household Level Production Function Using Weighted Ordinary Least Squares:[a] Log-Linear Function

Independent Variables	Corn	Peanuts	Millet	Sorghum
Labor	.589***[c]	.253**	.274**	-.016*
	(5.35)[d]	(1.50)	(2.05)	(-.24)
Capital Assets	.098	.148*	.014	.300***
	(2.57)	(1.84)	(.47)	(3.24)
Animal Traction (hours)	.141	.216**	.028	.086
	(1.96)	(2.20)	(.41)	(1.04)
Intercept	2.021	4.237	4.254	6.334
	N = 186	F = 11.858***	R^2 = .511	
Labor	.313***	.244***	.268***	.301***
	(3.91)	(3.22)	(3.87)	(4.43)
Capital Assets	.067*	.149*	.041	.242***
	(1.73)	(1.83)	(.45)	(2.57)
Animal Traction (hours)	.107**	.215**	.027	.105
	(1.45)	(2.21)	(.40)	(1.53)
Intercept	e	4.294[e]	e	e
		(10.07)[e]		
	N = 186	F = 12.419***	R^2 = .463	

[a]For detail of this method of estimation, see Wannacott and Wannacott (1979, 431-32).
[b]The dependent variable is logarithm of yield in kg/ha.
[c]Statistical significance levels: *** = 1%, ** = 5%, * = 10%.
[d]Figures in parentheses are t-values.
[e]Same intercept for all four crops. F-value for testing equal intercept for four major crop regressions is 1.7033 with degrees of freedom 3 (for numerator) and 170 (for denominator).

presented in Tables 6.8 through 6.10. The dependent variable in all the equations is the log of total production of the crop concerned. Estimates were obtained using several alternate specifications of the model to examine the consistency and stability of the estimated coefficients on the variables and the emerging relationships. As evident from the results presented in the tables, the coefficient of the animal traction hours variable appears consistently positive in all the estimated equations and for all the individual crops for which farmers used animal traction. The estimated coefficient of the variable is consistently large in size and statistically significant for the peanut and corn crops (Tables 6.8 and 6.9). For millet and sorghum crops also, the estimated coefficient of animal traction variable is positive, although statistically relatively weak in the per hectare yield functions (Tables 6.8 and 6.9). However, the effect of animal traction on total yield appears positive and statistically significant at the 5 percent level for sorghum while at the 15 percent level for millet (Table 6.10). The size of the animal traction effect on total sorghum output is much larger than the millet output.

The results of the estimated farm production relationships presented in Table 6.10 are important also from another aspect. The effect of animal traction on total output of the two major crops of the region has been analyzed by taking into account the effect of soil types[3] on crop yields (in Tables 6.8 and 6.9, the soil types were excluded from the equations). The results of the statistical analysis presented in Tables 6.8 through 6.10 reveal that the effect of animal traction on crop yields is positive and statistically strong in most cases. Additionally, the soil type effect on crop output also appears significant; the sandy soil (type) has a significantly positive effect on household's millet production, while in the case of sorghum, except for the clayey soil type, all the three soil types, i.e., sandy, gravelly and mixed show strong positive effects on output. An important thing to note in the results of Table 6.10 is that the inclusion of the soil type variable in the model, which is, in fact, an indicator of soil quality as seen by the farmers, resulted in making the effect of animal traction on crop output more visible and powerful

The results on the other variables such as labor, capital and farm size confirm those reported and discussed earlier in Chapter 5; hence, no further comment or discussion is warranted again in this section.

TABLE 6.10
Estimated Regression Coefficients of the Household Level Production Function Using Ordinary Least Squares (OLS) for Millet and Sorghum:[a] Log-Linear Function

Independent Variables	Millet	Sorghum
Yield of Crops Grown in Association	-.054 (1.26)[b]	.0796 (1.59)
Planting Labor (hours[d])	.298*[c] (1.85)	.302* (1.70)
Tilling Labor (hours[d])	.0661 (1.29)	.054 (.68)
1st Weeding Labor (hours[d])	.0457 (.25)	-.2 (1.05)
2nd Weeding Labor (hours[d])	-.058 (.66)	-.0772 (.088)
Use of Animal Traction (hours)	.062 (1.19)	.24** (2.5)
Input Expenses (in CFA francs)	.0101 (.39)	-.0615 (1.04)
Sandy Soil Area (hectares[e])	.158*** (4.05)	.564*** (3.35)
Gravelly Soil Area (hectares[e])	.038 (.879)	.345** (2.07)
Clayey Soil Area (hectares[e])	.148 (1.64)	.0378 (.245)
Mixed Soil Area (hectares[e])	.065 (.56)	.521** (2.13)
Constant 'a'	5.026*** (6.7)	5.471*** (9.32)
R^2	.605	.422
F	9.221***	4.45***
N (number of households)	60	53

[a]Dependent variable = total yield (kilograms).
[b]Numbers in parentheses are t-statistics.
[c]Statistical significance levels: *** = 1%, ** = 5%, * = 10%.
[d]All labor hours of the males, females and children working in the fields.
[e]One hectare = 2.47 acres. These soil types variables are measured in terms of the total land area under each of the individual soil type on which millet and sorghum were grown by households in the sample.

Results of the Field or Parcelle Level Crop Production Function

The availability of crop production and input use data by field or parcelle enabled this study to conduct the production function analysis also at a disaggregated (field) level and compare the results with those of the aggregate (household or farm) level analysis reported in Tables 6.8 through 6.10, and to examine the consistency and/or stability of the estimated relationships. The results of estimates obtained through several alternative specifications of the estimating field level production model are presented in Tables 6.11 through 6.14. To mention once again the fact that farmers in the region operate small farms which are divided into several (scattered) fields or plots (parcelles) some of which are located close to the village while others away from the village in the bush. One farmer, therefore, operates several parcelles of land; and these parcelles (fields) differ in terms of soil type (quality), size, the pattern of cropping, and input use. The results in Tables 6.11 through 6.14 are based upon the input-output data by individual fields or parcelles managed by households. However, the results of the estimating production function are extremely instructive. To state briefly, the effect of the size of the parcelle is positive on total output (Tables 6.11 and 6.12) while it is negative on per hectare output (Table 6.13). The labor input also shows a positive impact on output. The effect of purchased inputs appears mixed, and in most cases statistically it is not significant. As these variables have already been discussed earlier in Chapter 5, any further discussion on these variables will be simply repetitive.

The significant part of the results of estimates (Tables 6.11, 6.12, and 6.13) is with respect to the relationship between animal traction and crop output. First, consider the estimates presented in Table 6.11. The effect of soil types has been

TABLE 6.11
Estimated Coefficients of the Field or Parcelle Level Production Function Using Ordinary Least Squares for Millet and Sorghum: Log-Linear Function[a]

Independent Variables	Millet	Sorghum
Size of Parcelle	.628	.587
(hectare)	(10.44)[b]	(12.26)
Yield of Crops Grown	.00847	.0282
in Association	(.261)	(.64)
Planting Labor	.1413**[c]	.0875
(hours[d])	(2.36)	(.81)
Tilling Labor	.0052	.0272
(hours[d])	(.154)	(.48)
1st Weeding Labor	.0304	-.0074
(hours[d])	(.469)	(.07)
2nd Weeding Labor	.0543*	.066
(hours[d])	(1.85)	(1.36)
Use of Animal Traction	.0522	.221**
(hours[d])	(.98)	(2.03)
Input Expenses	.01514	-.111*
(CFA francs)	(.55)	(1.89)
Constant 'a'	4.855***	5.087***
	(17.52)	(12.26)
Adjusted R^2	.638	.481
F	70.13***	17.91***
N (number of fields)	314	147

[a]The dependent variable is the total yield of the crop (kg).
[b]The numbers in parentheses are t-statistics.
[c]Statistical significance levels: *** = 1%, ** = 5%, * = 10%.
[d]All labor hours contributed by males, females and children working in the fields.

ignored in these estimates--the soil type variables are excluded from the estimating equations for millet and sorghum production. However, the variables of the model in both cases explain 49 to 67 percent of the variations in crop output, and, as shown by F-statistics, the coefficient of multiple determination, R^2, is statistically highly significant (Tables 6.11 and 6.12).

As shown by the results,the effect of animal traction on sorghum and millet (total) production appears positive, although the coefficient is statistically significant only in the equation for sorghum production. However, the inclusion of soil type variable in the crop production model resulted in enhancing the effect of animal traction also for millet production. The value of the coefficient on animal traction hours increases from 0.005 in the millet equation without the soil type variable in the model (Table 6.11) to 0.09 in the equation with the soil type variable included in the model (Table 6.12); also the coefficient turns statistically significant at the 10 percent level in the presence of the soil type variable in the estimating model (Table 6.12). Also interesting are the results of the statistical analyses presented in Tables 6.13 and 6.14. The estimates in these tables are based upon the per hectare yields of the two crops as influenced by animal traction along with other factors. The results of these tables are quite similar to those of Tables 6.11 and 6.12 with respect to the effect of animal traction. The main thing to note, however, is that the use of animal traction on farm production is seen through its impact on farm productivity measured by the per hectare yield of the crop. The estimated coefficient of animal traction on per hectare productivity appears positive in both equations (Table 6.13) but it is statistically significant only in case of sorghum, a result similar to that of Table 6.11. Also note that the estimate of the production function for millet shows a positive and statistically significant effect of animal traction in the case of gravelly type soil (Table 6.14).

Overall, it is clear from the results of Tables 6.11 through 6.14 that the estimates seem quite stable and consistent in terms of the nature as well as the strength of the effect of the animal traction technology on small farm production system. The results demonstrate that in the present production setting animal traction has a positive influence on both (total) production and (per hectare) productivity of land under crops raised in the region; of course, the extent of the effect somewhat differs across the major crops. Furthermore, the inclusion of the soil type variable in the production model resulted in not only improving the estimated relationship

TABLE 6.12
Estimated Coefficients of the Field or Parcelle Level Using Ordinary Least Squares for Millet and Sorghum: Log-Linear Function[a]

Independent Variables	Millet	Sorghum
Size of Parcelle (hectare)	.6621***[b] (11.58)[c]	.577*** (6.51)
Yield of Crops Grown in Association	.0131 (.425)	.0287 (.24)
Planting Labor (hours[d])	.1356** (2.38)	.107 (.98)
Tilling Labor (hours[d])	.0142 (.443)	.0175 (.31)
1st Weeding Labor (hours[d])	.0363 (.58)	.0258 (.24)
2nd Weeding Labor (hours[d])	.0329 (1.2)	.0515 (1.05)
Animal Traction (hours)	.0901* (1.81)	.219** (2.02)
Input Expenses (CFA francs)	.0101 (.38)	-.115* (1.94)
Gravelly Soil (dummy[e])	-.5035*** (4.9)	.440* (1.814)
Clayey Soil (dummy[e])	-.1528 (1.05)	--
Sandy Soil (dummy[e])	--	.169 (.77)
Mixed Soil (dummy[e])	-.495** (2.01)	-.0516 (.20)
Constant 'a'	5.118*** (19.21)	4.775 (10.43)
R^2	.666	.488

TABLE 6.12 (Continued)

Independent Variables	Millet	Sorghum
F	59.9***	13.66***
N (Number of Fields)	326	147

[a]The dependent variable is total yield with soil type variable.
[b]Statistical significance levels: *** = 1%, ** = 5%, * = 10%.
[c]Figures in parentheses are t-values.
[d]All labor hours of the males, females and children working in the fields.
[e]For Millet the control soil type variable was sandy, and hence the other soil types dummies were defined as follows: gravelly dummy = 1 for gravelly soil and 0 for the rest; clayey dummy = 1 for clayey soils and 0 for the rest; mixed dummy = 1 for mixed soils and 0 for the rest.

For Sorghum the control soil type was clayey: gravelly dummy = 1 for gravelly soils and 0 for the rest; sandy dummy = 1 for sandy soils and 0 for the rest; mixed dummy = 1 for mixed soils and 0 for the rest.

between animal traction and crop production under a varied set of conditions (and specifications) imposed on the model but also revealing the effect of soil types themselves on crop production.

IV. SUMMARY AND CONCLUSION

The results of the descriptive analysis indicate that although the direction of causality between animal traction (AT) and farm size is not determined, the total land area cultivated by AT households is twice as large as that of the non-AT households. The land area cultivated per active worker is also, on the average, significantly greater on the animal traction farms than on the non-AT farms. Second, a unit labor hour of AT households cultivated about 30 percent more land area than a unit labor hour of non-AT households. Third, labor input per land area of the non-AT households is greater than that of the AT households for all labor operations and the disaggregated labor categories. This evidence supports the hypothesis that an

TABLE 6.13
Estimated Coefficients of the Field or Parcelle Level Production Function Using Ordinary Least Squares For Millet and Sorghum: Log-Linear Function[a]

Independent Variables	Millet	Sorghum
Size of Parcelle (hectare)	-.358***[b] (6.0)[c]	-.412*** (4.6)
Yield of Crops Grown in Association	.00136 (.04)	.0282 (.64)
Planting Labor (hours[d])	.138** (2.34)	.087 (.813)
Tilling Labor (hours[d])	.0128 (.39)	.027 (.48)
1st Weeding Labor (hours[d])	.0304 (.47)	-.0074 (.07)
2nd Weeding Labor (hours[d])	.039 (1.4)	.066 (1.36)
Input Expenses (CFA francs)	.0154 (.56)	-.111* (1.89)
Animal Traction (hours)	.067 (1.3)	.2212** (2.03)
Constant 'a'	4.912*** (18.0)	5.087*** (12.26)
R^2	.19	.143
F	5.17***	4.05***
N	326	147

[a]The dependent variable is per hectare yield (kg/ha).
[b]Statistical significance levels: *** = 1%, ** = 5%, * = 10%.
[c]Figures in parentheses are t-values.
[d]All labor hours of the males, females and children working in the field.

TABLE 6.14
Estimated Coefficients of the Field or Parcelle Level Production Function Using Ordinary Least Squares For Millet: Log-Linear Function by Soil Types[a]

Independent Variables	Sandy Soil	Gravelly Soil	Clayey Soil
Size of Parcelle (hectare)	.596***[b] (7.03)[c]	.4917*** (4.22)	.852 (5.8)
Yield of Crops Grown in Association	.034 (.7)	-.025 (.44)	-.0062 (.078)
Planting Labor (hours[d])	.132* (1.6)	.0274 (.257)	.1131 (.648)
Tilling Labor (hours[d])	.00103 (.024)	.0103 (.16)	.1119** (1.149)
1st Weeding Labor (hours[d])	.058 (.58)	.229** (2.0)	-.1391 (.97)
2nd Weeding Labor (hours[d])	.0769* (1.87)	.0679 (1.49)	.055 (.528)
Use of Animal Traction (hours[d])	.0109 (.136)	.1985** (2.605)	-.0515 (.29)
Input Expenses	.00316 (.08)	.011 (.236)	.035 (.43)
Constant 'a'	4.816*** (11.99)	3.797*** (6.82)	5.804*** (9.021)
R^2	.64	.616	.798
F	39.04***	21.45***	20.31***
N	171	103	40

[a]The dependent variable is total yield (kg).
[b]Statistical significance levels: *** = 1%, ** = 5%, * = 10%.
[c]Figures in parentheses are t-values.
[d]All labor hours of the males females and children working in the field.

animal traction program is a labor-saving technology. The use of animal traction saves labor input by 30 to 40 percent on the average. Fourth, the output-labor ratios per unit land of AT households are 2.5 times larger than those of non-AT households in cases of peanut and corn production, although in cases of millet and sorghum productions, no significant difference in the output-labor ratios was observed.

The results of the econometric analysis conducted through several alternative specifications of the estimating production functions show that the use of animal traction by small farmers results in improving the productivity of land; the estimated coefficient of animal traction shows a positive and statistically significant impact of the variable on total output as well as on per hectare yield on the farm. The extent of the effect of animal traction varies from crop to crop, however. The introduction of the soil type (reflecting land quality) variable in the production function analysis improved the estimated relationships; for example, as a result of the inclusion of the soil type variable in the estimating equations, the effect of animal traction on crop production emerged powerful both in the total production function and the per hectare yield function. The results remain stable and consistent across several alternative specifications of the estimating model.

It is clear from the results of this study that the use of animal traction has several aspects with important economic consequences for the small farmers. These relate to the land area farmed, saving in labor, and more significantly, improvement in crop (yield) productivity. The farmers have become aware of the economic benefits, in addition to the usual status symbol associated with the possession of animal traction (the plow and the draft animals to draw the plow or other implements--planters or seeders, for example). This is evident from the growing demand for animal traction in the region. However, there are problems and constraints facing the farmers. It is in this context that we need to focus on appropriate public policy directed towards the provision of credit, the supply of trained animals, improved farm implements, education and extension, in addition to the supply of other improved farm inputs to be made available to farmers. Furthermore, the World Bank's guidelines as well as the Bank's assistance in the promotion of the animal traction technology, indeed of the new agricultural technology in the African region of traditional, human labor-intensive farming systems, will assume significance. Above all, the nation state's own (internal) economic policies and priorities become pivotal in such developmental efforts that aim at improving the economic lot of the farm people.

NOTES

1. Based upon personal interviews conducted by the author covering approximately 40 farmers under a Credit Constraint survey. Questions addressed in the survey included, among other things, the nature and extent of production constraints faced by farmers. An important purpose for which credit was sought by farmers was said to be the acquisition of animal traction.

[2]For details, see Wannacott and Wannacott (1979, 431-32). Also, several other references on agricultural production functions listed in the References section of the chapter are recommended.

[3]Soil types have been classified into 4 broad categories (i.e., soils which are mostly sandy, or gravelly, or clayey, or mixed). These categories are based upon the results of a survey on soils conducted in the region as a part of the Farming Systems Research Study. The soils were classified by the farmers studied in the sample. It needs to be stated that the farmers' classifications and observations are based upon their past experience, their knowledge and information about the different kinds of soil characteristics (or types) and their suitability for growing different crops that have been passed on over time from one generation to another.

The farmers in the sample, including some chiefs of villages, had identified 13 types of soil characteristics or types of soil depending upon their suitabilities for growing several varieties of crops. The types of soil identified by farmers and the criteria used by them appeared reasonable and indeed meaningful from the standpoint of crop patterns followed in the region. Important considerations taken into account by farmers were whether the soil was sandy, gravelly, clayey or mixed, the water retention capacity of the soil, and the location of the fields from the house compound.

The 13 classifications identified by the sample farmers and the village chiefs were later regrouped into 4 major types after consulting local research supervisors and some agronomists and soil experts. However, it needs to be noted that these types are based upon the farmers' classifications of soils. It may be useful to provide some major statistics, as shown in the table

Crops and Soil Types[a]

Crops and Soil Types	Total Area (hectares)	%	Total No of Parcelles	Per-Hectare Yield (kg)
Millet:				
Sandy Soil	98.68	47	171	575
Gravelly	70.59	34	103	381
Clayey	27.24	13	40	550
Mixed	11.29	6	12	409
Sorghum				
Sandy Soil	22.79	36	56	680
Gravelly	14.07	22	35	788
Clayey	16.02	25	29	399
Mixed	11.04	17	27	481

[a]These estimates are based upon the sample survey data for 1979-1980.

above pertaining to soil-types as classified by farmers, area allocated to different soil types by crops, and the per hectare yields.

REFERENCES

Barrett, V., G. Lassiter, D. Wilcock, D. Baker, and E. Crawford. *Animal Traction in Eastern Upper Volta: A Technical, Economic and Institutional Analysis*. East Lansing, Michigan: Michigan State University, Dept. of Agr. Econ., MSU IDP No. 4, 1982.

Christensen, L.R., D.W. Jorgenson, and L.J. Lau. "Transcendental Logarithmic Production Frontiers." *Review of Economics and Statistics* 55(1973):23-45.

Delgado, Christopher and J. McIntire. "Constraints on Oxen Cultivation in the Sahel." *American Agricultural Economic Journal* 64(1982):188-96.

Eicher, Carl K. and Doyle C. Baker. *Research on Agricultural Development in Sub-Saharan Africa: A Critical Survey*. East Lansing, Michigan: Michigan State University, Dept. of Agr. Econ., MSU IDP No. 1, 1982.

Ferguson, C.E. *The Neoclassical Theory of Production and Distribution*. New York: Cambridge University Press, 1969.

Griliches, Z. *Production Functions in Manufacturing: Some Preliminary Results*. Amsterdam: North-Holland Publishing Company, 1967.

Griliches, Z. and V. Ringstad. *Economies of Scale and the Form of the Production Function*. Amsterdam: North-Holland Publishing Company, 1971.

Hayami, Y. and V.W. Ruttan. *Agricultural Development: An International Perspective*. Baltimore: Johns Hopkins University Press, 1971.

Heady, E.O. and J.L. Dillon. *Agricultural Production Functions*. Ames, Iowa: Iowa State University Press, 1961.

Henderson, J.M. and R.E. Quandt. *Microeconomic Theory: A Mathematical Approach*, 2nd ed. New York: McGraw-Hill, 1971.

Intrilligator, M.D. *Econometric Models, Techniques & Applications*. Englewood Cliffs, New Jersey: Prentice-Hall, 1978.

Jaeger, William K. *Animal Traction and Resource Productivity: Evidence from Upper Volta*. SAFGRAD Program, FSU Project, Purdue University, 1984.

Johanson, M. *Production Functions*. Amsterdam: North-Holland Publishing Company, 1972.

Johnston, J. *Econometric Methods*. New York: McGraw-Hill, 1972.

Lang, M., R. Cantrell and J. Sanders. *Identifying Farm Level Constraints and Evaluating New Technology in Upper Volta*. SAFGRAD Program, FSU Project, Purdue University, 1983.

Lele, U. "Rural Africa: Modernization, Equity and Long-Term Development." *Science* 211(1981):547-53.

Lunning, H.A. *Economic Aspects of Low Labor-Income Farming*. Wageningen, Netherlands: Centre for Agricultural Publications and Documentation, Agricultural Research Report, No. 699, 1967.

Mayor, T.H. "Some Theoretical Difficulties in the Estimation of the Elasticity of Substitution from Cross-Section Data." *Western Economic Journal* 7(1969):153-63.

McFadden, D. "Further Results on CES Production Functions." *Rev. Econ. Studies* 30(1963):73-83.

Mundlak, Y. "Elasticities of Substitution and the Theory of Derived Demand." *Rev. Econ. Studies* 35(1968):225-36.

Nerlove, M. *Estimation and Identification of Cobb-Douglas Production Functions*. Amsterdam: North-Holland Publishing Company, 1965.

Normal, D.W., D.H. Pryor, and C.J.N. Gibbs. *Technical Change and the Small Farmer in Hausaland, Northern Nigeria*. East Lansing, Michigan: Michigan State University, Dept. of Agr. Econ., African Rural Economy Paper No. 21, 1979.

Ruthenberg, H. *Smallholder Farming and Smallholder Development in Tanzania: Ten Case Studies*. Munich, West Germany: Weltforum Verlag, 1968.

Sargent, M., J. Lichte, P. Matlon, and R. Bloom. *An Assessment of Animal Traction in Francophone West Africa*. East Lansing, Michigan: Michigan State University, Dept. of Agr. Econ., African Rural Economy Working Paper No. 34, 1981.

Sato, K. *Production Functions and Aggregation*. Amsterdam: North-Holland Publishing Company, 1975.

Saylor, R.G. "Farm Level Cotton Yields and the Research and Extension Services in Tanzania." *Eastern Africa Journal of Rural Development* 7(1974):46-60.

Schultz, T.W. *Transforming Traditional Agriculture*. New Haven: Yale University Press, 1964.

Shapiro, K.H. *Efficiency and Modernization in African Agriculture: A Case Study in Geita District, Tanzania*. Ph.D. Dissertation, Stanford University, 1973.

__________. "Sources of Technical Efficiency: The Roles of Modernization and Information." *Econ. Dev. and Cultural Change* 25(1977):293-310.

__________. "Water, Women and Development in Tanzania." Paper presented at the Third Annual Conference of the International Water Resources Association, Sao Paulo, Brazil, 1978.

Singh, Ram D. *Small Farm Production Systems and Their Relevance to Research and Development: Lessons from Upper Volta, W. Africa*. Chicago: University of Chicago, Dept. of Economics, Workshop Paper No. 81:20, 1981.

__________. *Major Cropping Patterns of SAFGRAD Countries, Upper Volta: Facts and Observations to Farming System Research*. Ibadan, Nigeria: SAFGRAD, Document No. 7, 1981.

Singh, R.D., E.W. Kehrberg and W.H.M. Morris. *Small Farm Production Systems in Upper Volta: Descriptive and Production Function Analysis*. Station Bulletin No. 442, Dept. of Agr. Econ., Purdue University, 1984.

Spencer, D.S.C. and D. Byerlee. "Technical Change, Labor Use, and Small Farmer Development: Evidence from Sierra Leone." *American Journal of Agricultural Economics* 58(1976):874-80.

Tench, A.B. *Socio-Economic Factors Influencing Agricultural Output: With Special Reference to Zambia*. Saarbrucken: Verlag des SSIP-Schriften, 1975.

Walters, A.A. "Production and Cost Functions: An Econometric Survey." *Econometrica* 31(1963):1-66.

Wannacott, R.J. and T.H. Wannacott. *Econometrics*, 2nd ed. New York: John Wiley & Sons, 1979.

Welsch, D.E. "Response to Economic Incentives by Abakaliki Rice Farmers in Eastern Nigeria." *Journal of Farm Economics* 47(1965):900-14.

Wharton, Clifton R., Jr. *Subsistence Agriculture and Economic Development*. Chicago: Aldine Publishing Company, 1969.

The World Bank. *Accelerated Development in Sub-Saharan Africa: An agenda for Action*. Washington, D.C., The World Bank, 1981.

__________. The World Bank Development Report. Washington, D.C., The World Bank, 1984.

Zellner, A., J. Kmenta and J. Dreze. "Specification and Estimation of Cobb-Douglas Production Function Models." *Econometrica* 34(1966):727-9.

7

Agricultural Research in the Region and New Farm Technology: Adoption Problems and Prospects

I. AGRICULTURAL RESEARCH IN THE REGION

Agricultural research in the region has been carried out over several decades by the Institute for Research in Tropical Agriculture (IRAT), a French-supported institute, followed by the United Nations Development Project (UNDP) supported International Crop Research Institute for Semi-Arid Tropics (ICRISAT), set up in the country in the early seventies, and, lately, by the U.S. Agency for International Development (USAID) supported Farming Systems Research which initiated a multi-disciplinary research effort in 1979. Interesting, although very limited, yield results from the experimental varietal trials have been reported by agricultural scientists working at these centers. Compare, for example, the existing farm yield levels reported in Chapter 5 (Tables 5.7 and 5.14), with those realized at the experiment stations, the research managed trials, and the model farms presented in Table 7.1 of this chapter. In particular consider the case of sorghum and maize, for which data are available. The ICRISAT's (1980-1981) sorghum variety E-35-1 has the potential of an average yield of 3.5 to 4 metric tons per hectare with the recommended fertilizer applications and management practices. Such yield levels are, of course, realized under highly controlled conditions which are currently difficult to attain in farmers' fields.

The (Purdue University's) Farming Systems Research Unit's managed farmer-field trials conducted during 1980 (Table 7.2) yielded 1.8 metric tons of grain per hectare of E-35-1, and 1.3 metric tons of SVP 35, the two sorghum varieties which were said to be promising for semi-arid regions in Africa. It needs to be emphasized that more evidence is needed to evaluate the performance of these varieties under actual farm conditions.

TABLE 7.1
Yields of Sorghum and Maize Realized at Experiment Stations and Model Farms, Burkina Faso, 1979-1980

Variety	Average Yield (kg/ha)	Observations
Sorghum	3,500 to 4,000	Reported by scientists of ICRISAT on the basis of experimental results.
Red Sorghum Saria Model Farm (IRAT) 1969-1974	2,551	IRAT's model farm in Saria (HV) with 4.4 hectares of cropland since 1969, with 6 persons (3 active), was phased to bring 1 hectare a year under improved technology. Yield figures arrived in fifth year.
IRAT P & K	975	0 level
Experiments for sorghum	1,806	50 kg of P205/ha (16.6 kg grain/kg of P205)
1964-1974	1,958	100 kg P205/ha (3 kg grain/kg P205)
	1,228	0 level
	1,679	50 kg K20/ha (9 kg of grain/kg K20)
	1,846	100 kg K20/ha (3.4 of grain/kg K20)
Maize		
IRAT 100	3,023	Mean yield based on IITA's trials in Burkina Faso, Senegal, Mali, Ivory Coast & Benin (1979)
B D S III	2,970	Mean yield based on IITA's trials in
Jaune de Fo	2,328	Burkina Faso, Senegal, Mali, Ivory
Massayomba	2,286	Coast & Benin (1979) 3 years average
Cowpea	1,500	based on IITA SAFGRAD trials

Source: ICRISAT, IITA/SAFGRAD and IRAT, *Reports*, 1979 and 1980.

TABLE 7.2
Yields of New Sorghum and Cowpea Varieties on FSU/SAFGRAD Research Managed Fields in Sample Villages, Burkina Faso, 1980

Crop Variety	Yield (kg/ha)	Observations
E-35-1 (Sorghum)	1,800	Village fields with preplanting cultivation and 100 kg RP + 20 kg urea per hectare. 1980 research-managed trial.
	1,500	Village fields without preplanting cultivation: no fertilizer. 1980 research-managed trial.
	750	Bush fields with preplanting cultivation and 1,200 kg RP + 20 kg urea per hectare. 1980 research-managed trail.
	150	Bush fields without preplanting cultivation: no fertilizer. 1980 research-managed trial.
SVP 35 (Sorghum)	1,300	Sandy valley soils Ouahigouya 1980 with preplanting cultivation and 100 kg rock phosphorous and 20 kg urea per hectare. Research-managed trial.
	600	Sandy valley soils Ouahigouya 1980 without preplanting cultivation, no fertilizer. Research-managed trial.

Source: Farming Systems Research Field Trials in Sample Villages, 1979-80.

Under the usual farm practices and soil fertility levels, E-35-1 did not appear to yield more than local varieties. More than a simple change of varieties may be involved if higher yields at the farm level are to be achieved.

On IRAT's experimental plots and Saria (research station) model farms (Table 7.1), the average per hectare yield of sorghum ranged between 2 to 2.5 metric tons. The model farm technology in IRAT's case (1979, 1980, 1981) was highly controlled and subsidized. The recipients of this subsidized technology were the employees of the Institute that was diffusing the new technology.

Maize yields reported by maize agronomists and breeders (IITA, IRAT) vary from over 2 metric tons to over 3 metric tons per hectare. Such yield levels are related to different levels of fertilizer applications, management practices, and varietal changes under West African farming conditions. The feasibility of realizing the yield potentialities of the new varieties under farm conditions has yet to be established.

It is unlikely that the ideal or potential yield of 3.5 to 4 metric tons of grain per hectare will be achieved under farm conditions for either sorghum or for maize. Of course, there may be areas and farmers with relatively more favorable conditions for which yields higher than 1 to 2 metric tons per hectare are fairly attainable. The national average yield statistics for cereals is only about 500 kg per hectare, a figure which hides yield potentialities in the regional and subregional contexts. For example, as shown by the data in Table 5.7 (chapter 5), the average yield for sorghum ranges from as low as 148 kg per hectare in the Dori region and 368 in the Yatenga region to as high as 844 kg per hectare in the Bobo region and 848 in the Fada region. Likewise for maize, it ranges from 206 kg per hectare in Yatenga to 1,230 in Fada and 1,045 in Bobo. For AVV[1] farms, it is estimated to be over 1,000 kg per hectare. For other crops there is a similar pattern of yield differentials.

Such productivity differences in existing farming systems in the country may give some useful guidelines for comparing the experiment stations' yields with the existing yields already realized by farmers in different regions, especially by those who are already obtaining yields around 1 metric ton or more per hectare. Based on only four paired comparisons in one village, the Purdue University's Farming Systems Research Unit reported the mean yield of E-35-1 (sorghum) at 1,120 kg per hectare as compared to the local sorghum yield of 1,690 kg per hectare with the same input usage. However, other observations of E-35-1 in the same area, but unfortunately with no local checks showed an average yield of 1,720 kg per hectare. In this case, the two crop varieties were planted on relatively high quality village fields.

It is possible that some of the local varieties may yield as much as new (or improved) varieties do under similar conditions of management practices and input use. In such cases the farmers will have little incentive to try the new variety. The relative superiority of any new technology has to be clearly demonstrated. For valid comparisons and meaningful extrapolations, the macro level average yields do not represent a true picture. It is necessary to compare yields in the regional, subregional and even village context. This can identify regions with different potentialities with respect to various crop varieties, cropping systems and crop improvement programs. If the objective is to achieve maximum increase in cereal production in as short a period as possible for countries such as Burkina Faso, scarce research and development resources need to be allocated on selective bases with relatively higher priorities for areas with greater potential for using yield-increasing technologies, and higher economic returns to investment.

II. LOW FARM YIELDS, NEW TECHNOLOGY, ADOPTION PROBLEMS AND PROSPECTS

The farm level yield data collected in the present study indicate extremely low crop yield conditions in the agriculture of the region. There are several questions and issues involved in a search for the variables causing low yields and for the ways to effectively promote yield-increasing technology. One is: Do we, more specifically those in charge of planning and development efforts, have adequate information and understanding to create conditions for improving farming practices enough to increase yields per hectare from the present low level of around 500 kg per hectare to, say, 1,000?

Despite the fact that some of the improved varieties of crops such as sorghum and maize have been found to give much higher yields, 3 to 4 metric tons of grain per hectare on experimental plots at research stations under highly controlled conditions, and despite the successful performance of these new crop varieties, and other "improved" practices in some farmers' fields, they have not been fully accepted. If the farmers were informed and convinced that the new maize varieties perform profitably under their conditions and constraints, adoption of such technology could be expected.

Apparently the farmers are not convinced that the new varieties maximize returns to their scarce resources. An irregular supply of modern inputs at affordable prices may affect the situation. Additionally, the new varieties may differ from the traditional varieties with respect to timing of labor requirements. This may create labor constraints that affect timeliness in planting and in performance of other field operations that are sensitive to rainfall patterns.

On the other hand, sorghum, millet, and maize prices have risen about twice as much since 1968 as the amount of inflation and the prices of alternative crops.[2] This should favor higher yielding varieties unless the prices are offset by an unfavorable input cost situation.

Grain marketing conditions probably do not favor technological improvement and growth in the farm productivity of Burkina Faso; and this condition holds for several other countries in the region. From 1978 to 1980, for example, the official government prices in Burkina Faso ranged from between 40 and 45 CFA francs per kg of millet and 32 to 37 CFA francs per kg of sorghum.[3] These prices were lower than the open market prices as evident from the fact that in the latter half of 1980, farmers were selling millet and sorghum in open markets for 60 to 75 CFA francs per kg. At that time the government was considering setting the minimum prices[4] of 40 to 45 CFA francs per kg. Even though this price policy provides some disaster insurance, it may not alter greatly the farmers' view of the risk associated with investments required to increase the production of the various crops. The farmers do what they consider most advantageous under the local marketing conditions, but there may be little incentive to take any risks with modern inputs that may or may not pay off immediately under the variable rainfed production conditions coupled with a relatively uncertain product price situation.

In low resource areas of high risk farming new technology has to be low in cost if the individual farmers are to adopt it without special incentives and outside assistance. Farmers in the plateau area of the region run high risks of crop losses from lack of rainfall and its erratic distribution. They have no control over this variable and risk of crop losses cannot be completely eliminated with rain-fed farming in semi-arid zones of Africa. A great challenge to agricultural scientists is to evolve technologies that permit crop production to better withstand these weather conditions, technologies that fit well into the known farming systems, and insure higher return to farmers than the traditional technologies that have evolved by trial and error.

Other conditions faced by farmers in the study region are equally unfavorable. In general the farm extension services do not reach most of the farmers. These services are poorly organized, lack trained personnel, and have limited financial resources. In one of the five sample villages, farmers said they had not seen any ORD extension agent for the last ten years! In a country where 95 to 98 percent of the farm population is illiterate, a weak and often inefficient system for extension of technical information can seriously limit technological change and improvement in agriculture. An innovation does not spread unless there are effective communication linkages.

Availability of input supplies such as chemical fertilizers, insecticides and pesticides, farm equipment, draft animals to pull such equipment, and the lack of credit to buy these modern inputs pose serious problems to farmers. Poor transportation coupled with inefficient input markets can cause farmers to view modern input use an uneconomical. This may further discourage farmers from investing in new production technology.

It is extremely difficult to increase agricultural production under present conditions. There are strong forces favoring the status quo in the production system. It may very well be an efficient agriculture under the existing conditions and constraints. However, it is not a progressive, moving agriculture from the viewpoint of the needs of the country in a changing world.

This situation does not imply that farmers in Africa are primitive, backward, inefficient and irrational because they follow old production practices. Their action merely indicates rational resource allocation decisions and choice of production practices under the set of conditions and constraints with which they are faced. The view that these farmers are irrationally following outdated practices indicates (a) a failure to appreciate the social and economic realities of farming in these regions, and (b) a failure to understand the constraints of the existing farming systems.

Micro Level Crop Substitution: A Comparative Perspective, and Problems[5]

Sorghum and maize are two food grain crops for which improved varieties are being promoted for farmers in Burkina Faso and in other parts of West Africa. In most cases these crops compete with each other for fertile soils. However, farmers generally allocate a relatively greater proportion of cultivated land to sorghum than to maize. Maize is relatively more sensitive to weather conditions than is sorghum, and farmers run greater risks of losing this crop when there is drought. As a rainfed crop sorghum has relatively greater probability of survival under drought conditions than the maize, other things equal.

The relative cost-benefit perspective of sorghum and maize can be altered by technological changes, such as introduction of a drought resistant high yielding seed variety, or modification of the existing management practices. If this makes maize relatively more profitable, the chances for allocating more land to this crop will increase. Effective profitability may be realized through lessening risk of low yields, reducing per unit cost of production, or increasing per hectare yield with the same input cost. Since maize occupies a relatively much smaller fraction of total cultivated land under the existing farming system, one might expect the area devoted to it to be potentially expandable.

However, maize production cannot be expanded over all of the country. The agro-climatic requirements of this crop give certain areas such as the Southwest and the Fada regions higher potential for production increase than the Central and Northern regions. These regions have a comparative advantage in terms of soils, rainfall and other favorable resource endowments. In the region of study, the maize area continues to be very small despite the fact that the marginal value product for land sufficiently fertile to support maize is much higher than the less fertile land needed for sorghum. Farmers have some constraints preventing them from expanding area under maize. The amount of fertile land available for maize including the cost of fertilizer and possibly the higher risks that maize may be subjected to as a result of inadequate and fluctuating rainfall are some of the constraints. In addition, a currently limited market for maize for roasting ears may be another constraint on expansion of maize area and production. The price for the dry grain is apparently not high enough to encourage expansion of this crop.

Let us take another case: the case of cowpeas vis-a-vis its competitors. In terms of kilograms of grain per hectare, cowpeas could compete with peanuts for land and other inputs if farmers were to plant it as a single crop. The improved cowpea variety, KN-1, performs much better when grown as a single crop since that permits certain necessary operations, particularly spraying, which are key elements influencing yield.

The yield potential for the cowpea variety KN-1 is at least 1,500 kg per hectare, given three to four sprayings of insecticides. Even if one assumes that in farmers' fields, the per hectare yield only reaches 1,000 kg, this would be much higher than the present yield levels of 200 to 250 kg per hectare.

A rough and quick cost benefit calculation suggests the following. For a hectare of land under the new cowpea variety, the farmer will need to incur a total cost of 31,800 CFA francs, or about 150 US dollars (labor cost = 11,600 CFA francs, seed cost = 3,200 CFA francs, about 14 US dollars, fertilizer 3,000 CFA francs, 13 US dollars, and spraying including variable costs and depreciation on the sprayer 14,000 CFA francs, 62 US dollars). He will receive a total revenue of 45,000 CFA francs, 200 US dollars (based on 45 CFA francs, 0.20 US dollars/kg and a yield of 1,000 kg/hectare under the new variety). The net revenue realized by the farmer will be 13,200 CFA francs, about 59 US dollars per hectare. However, to realize this net revenue, he will need to make an initial investment of 24,000 CFA francs, about 107 US dollars (of this the sprayer at the subsidized rate will cost 15,000 CFA francs, about 67 US dollars). Before a farmer makes any decision, he faces two important questions. First, how and where to get US $107 to undertake the initial investment. For small farm poor households this is not a small amount! Second, even if he were successful in getting the money, is it more advantageous for him to invest this money in the sprayer than elsewhere, e.g., to buy a houe manga or a donkey.

There are other questions as well. Will cowpea yield higher revenue per hectare of land than its competitor, other things being equal? We do not know if this is so. Marketing and pricing of cowpea, if production in the region changed, are other issues that would need consideration. Then there is the question of an infrastructure that would promote cowpea production. Farmers' knowledge and capabilities are essential elements in the whole process of spreading the new cowpea technology on small farms, knowledge about the use of sprayers to make them economical, money to buy the equipment, repair facilities, etc.

Various questions pertinent to new cowpea technology that need to be investigated include the following: (1) the relative profitability or returns from cowpea vis-a-vis its competitors; (2) the extent of competition for land and other resources among crops, e.g., cowpea and peanut; (3) the economic returns to sprayings--estimates of yield in relation to the timing and number of sprayings, and alternate uses of sprayers that make investment remunerative; and (4) the relative economics of cowpea production as a single versus an associated crop.

Conditions for Adoption of Improved Production Technology

Innovation is generally considered an important part of progress in a productive agricultural sector. A number of factors may affect the search for productive innovations and their acceptance by farmers. This section focuses on conditions for adoption of a different technology assuming that it is available for consideration.

1. The first and a necessary condition is that a new production technology has to be less costly in the use of the farmer's resources, or it achieves greater return from the same resources when compared with the traditional technology given the resource availability and constraints at farm level. Yield maximization per se is generally not one of the farmer's goals. His production decisions and choices involve trade-offs among goals. Under rainfed farming conditions he tries to reduce risks to an acceptable level. Those methods and technologies that increase achievement of one objective with little decrease in others given the farmer's economically most scarce resource (labor in many cases) are the best candidates for adoption.

2. A second condition is that the farmers have the knowledge and wisdom to evaluate the benefits and costs of the new technology and the skill or means of acquiring the skill to implement it. Involved here are farmers' training, schooling, the extension services, and other information systems. In the long run this means education of children and women which is a long-term investment. For payoff in the short term, adult education through extension services or other means is a likely necessity.

3. Improved infrastructure to serve rural areas, such as better roads and marketing facilities, is another condition that promotes movement of goods and services, information and people. The flow of technical information and information which facilitates exchange and efficient marketing as well as lower transport costs increases output-input price ratios which in turn furnishes incentive for economic changes.

4. Adequate input supplies, credit and distributional systems are needed to support yield increasing technologies. The current situation is inadequate with respect to both availability and stability of supplies and credit.

5. Farmers use expectations of market price to make decisions regarding levels of production, methods of production and product mix. This is especially true as economic growth occurs. Subsistence farmers may at first only sell surplus

crops, and the surplus is a very small fraction (10 to 15%) of their total production, but eventually as economic development proceeds, they will tend to sell more and seek more of the benefits of exchange. The pricing system should provide adequate incentives and give the producers adequate signals of the society's needs. For this to happen, governmental policy and actions must be consistent with those same needs and the government must be strong enough and stable enough to create a suitable political and economic environment. All sorts of market and price distortions must be removed.

Under the above conditions the farmer would be motivated to adopt improved farming practices and modify farming systems to achieve his goals and those of society. When appropriate and transferable technology is available under the above conditions, the farmer would have the incentive to use it. However, the farmer is generally expected to be shrewd enough not to accept any new idea until its benefit to him has been amply demonstrated.

III. SOME COMMENTS ON NATIONAL AND INTERNATIONAL RESEARCH AND DEVELOPMENT EFFORTS

There are several international institutions in the region(s) with competent and devoted agricultural scientists: plant breeders, agronomists, soil and water management specialists, entomologists, plant pathologists, and farming systems experts. Effective coordination is made rather difficult by the lack of a strong national research system that can furnish an appropriate linkage to avoid duplication, to promote areas of research best suited to the country's felt needs and priorities, and to monitor the flow of foreign aid in the area of agricultural research. This situation (namely, the lack of a strong national research system) is not uncommon in developing countries, particularly in the African developing countries.

It seems reasonable that a fairly substantial part of foreign assistance, no matter whether from individual countries or from international organizations, should be devoted to building strong national research capabilities with indigenous trained scientists. The initiative for this effort will have to come from the host country. A cadre of scientists and other experts subject to the vagary of foreign interests and funds cannot substitute for national scientists whose future depends on the host country. However, building and strengthening national research capabilities will require investment of resources in local research and educational institutions and facilities, training of local personnel, and making appropriate modification in the existing systems of education, and research. International educational and research organizations and their scientists also can be an important means of initiating such changes. These organizations can aid in the development of centers of higher learning and research (universities, colleges, research institutes) in the host country. There is certainly effort made to do this but, the fact is that still much more is required. In most cases, this will require additional funds for faculty and graduate students at the local institutions as well as for foreign experts. Such a process may be slow, but should be effective for developing indigenous capabilities in the long run.

Good working relationships and interaction among the various groups of international scientists are important. If they can use their limited resources to work harmoniously together to coordinate their research and avoid duplication they will be of better service to the host country or countries. Some of their resources allocated to areas of high pay off even when not the most popular projects can benefit the host country greatly. Having international scientists in the country may contribute to the host country's prestige internationally and make it difficult not to accept offers of all kinds of research. The donor countries must accept considerable responsibility for direction and coordination in this situation.

Formal schooling has been neglected in this country and several other less developed African countries. Educating rural people and farmers, men and women, and children may well be an investment with a very high payoff. Not more than 15 to 20 percent of rural children in Burkina Faso attend any kind of school. Illiteracy among farm women is almost one hundred percent. So far most of foreign aid received by the country has gone to the construction of physical capital rather than human capital. Foreign aid could play an important role in the creation of human capital in the farm population, the most neglected segment of the country's population. It is indeed sad that the farm population has been given least priority in the allocation of both national and international resources. Foreign aid for higher level training is also important, but training a few students at the graduate level is not enough. A broad-based foundation of human capital, the quality of human beings at the farm and community level, is essential to farm modernization. However, this requires some change in priorities of both the donor country and the receiving country with respect to foreign assistance.

It may be time that those of us concerned with agricultural development and welfare of the farm people in the developing countries pay heed to what Prof. T.W. Schultz said at Stockholm (Sweden) when delivering his Nobel lecture (1979) entitled "The Economics of Being Poor." To quote:

> We have learned that agriculture in many countries has the potential economic capacity to produce enough food for the still growing population and in so doing can improve significantly the income and welfare of the poor people. The decisive factors of production in improving the welfare of poor people are not space, energy and crop land. The decisive factor is the improvement in population quality.

The above comments are not meant to minimize the importance of previously stated observations on the need for efficient markets with appropriate price signals and incentives (and less governmental interventions and distortions in the market), construction of infrastructure necessary to sustain price signals, construction of infrastructure necessary to sustain a productive agricultural technology, and the establishment of effective backward and forward linkages between research and extension agencies, and priorities for allocating resources for agricultural development and research. These are important considerations with policy implications that must not be neglected if the development of the agricultural sector and the economy as a whole is desired.

To end the discussion on agriculture and farm production systems prevailing in the Western Sub-Saharan region, we may once again emphasize the following. It is possible to change varietal characteristics through breeding and experimentation. Although cost-benefit analyses indicate great potential gains, the problems of fitting the experimental varieties to the actual farming environment is complex, and the constraints facing the farmers need to be recognized by all the concerned, whether engaged in research, development, or policy. Furthermore, assistance programs that contribute to training local personnel as well as modifying the existing systems of education and research seem desirable for lasting impact. Similarly, considering that illiteracy rates are exceptionally high among the men, women and children engaged in the small-farm peasant agriculture of the region, perhaps the most under-rated technical assistance investment is simply formal schooling of rural people.

NOTES

1. This is a land resettlement organization under government control and supervision under which farmers are allocated land with a package of practices to be followed for different crops (in the Volta river basin).

2. See 1-8, World Bank, "Upper Volta Agricultural Issues Study," Report No. 3296-UV, October 29, 1982.

3. One US$ = 225-250 CFA francs.

4. See also note 8 in Chapter 5 of this book.

5. For a more complete land use discussion see Mahlon Lang, Ronald Cantrell, and John Sanders, "Identifying Farm Level Constraints and Evaluating New technology in the Purdue Farming Systems Project in Upper Volta." Paper presented at Farming Systems Symposium, Kansas State University, Manhattan, Kansas, October, 31, 1983.

REFERENCES

Christenson, P. Farming System Unit (FSU) Field Trials in Sample Villages. SAFGRAD/FSU Report, 1979-80.

Government of Upper Volta. *Annual Agricultural Statistics*. Ouagadougou: 1978-79.

Institute for Research in Tropical Agriculture (IRAT). *Annual Reports*. 1979, 1980, 1981.

International Crop Research Institute for Semi-Arid Tropics (ICRISAT). *Annual Reports*. Ouagadougou: ICRISAT, 1980-81.

International Fertilizer Development Center (IFDC). "West-Africa Fertilizer Study, Upper Volta." Technical Bulletin IFDC T-6, Vol. IV, March 1977.

International Institute for Tropical Agriculture (IITA). *Annual Reports*. Lagos, Nigeria: IITA, 1979, 1980.

Lang, Mahlon, R. Cantrell, and J. Sanders. "Identifying Farm Level Constraints and Evaluating New Technology in the Purdue Farming Systems Project in Upper Volta." Paper presented at Farming Systems Symposium, Kansas State University, Manhattan, Kansas, October 31, 1983.

SAFGRAD/FSU. *1982 Annual Report*. West Lafayette, Indiana: IE & R, IPIA, and Purdue University, 1983.

Saunders, Margaret O. "The Mossi Farming System of Upper Volta." FSU Working Paper No. 3, OUA/CSTR, Joint Project 31 between USAID and Purdue University, April 1980.

Schultz, T.W. "The Economics of Being Poor." Nobel Lecture, The Nobel Foundation, Stockholm, Sweden, December 10, 1979.

____________. *Transforming Traditional Agriculture.* New Haven & London: Yale University Press, 1964.

Singh, Ram D. "Major Cropping Patterns in SAFGRAD Countries, Upper Volta." Document #7, Ouagadougou, SAFGRAD/FSU, 1981.

Singh, Ram D. "Small Farm Production Systems in West Africa and Their Relevance to Research and Development." Agricultural Economics Workshop Paper No. 81:20, The University of Chicago, Department of Economics, May 28, 1981.

World Bank. *The World Development Reports.* Washington D.C.: World Bank, 1981, 1982, 1983, and 1984.

____________. *Accelerated Development in Sub-Saharan Africa: An Agenda for Action.* Washington D.C.: World Bank, 1981.

Index